HOW TO
SELL A
POISON

ELENA CONIS

HOW TO SELL A POISON

THE RISE, FALL, AND TOXIC RETURN OF DDT

BOLD TYPE BOOKS
New York

Bold Type Books
116 East 16th Street, 8th Floor, New York, NY 10003
www.boldtypebooks.org
@BoldTypeBooks

Printed in the United States of America

First Edition: April 2022

Published by Bold Type Books, an imprint of Perseus Books, LLC, a subsidiary
of Hachette Book Group, Inc. Bold Type Books is a co-publishing venture of the
Type Media Center and Perseus Books.

The Hachette Speakers Bureau provides a wide range of authors for speaking
events. To find out more, go to www.hachettespeakersbureau.com or call
(866) 376-6591.

The publisher is not responsible for websites (or their content) that are not
owned by the publisher.

Print book interior design by Jeff Williams.

Library of Congress Cataloging-in-Publication Data

Names: Conis, Elena, author.
Title: How to sell a poison : the rise, fall, and toxic return of DDT / Elena Conis.
Description: First edition. | New York : Bold Type Books, 2022. | Includes
    bibliographical references and index.
Identifiers: LCCN 2021047425 | ISBN 9781645036746 (hardcover) | ISBN
    9781645036753 (ebook)
Subjects: LCSH: DDT (Insecticide)—Toxicology. | DDT (Insecticide)—Health
    aspects. | DDT (Insecticide)—Environmental aspects. | DDT (Insecticide)—
    Physiological effect.
Classification: LCC RA1242.D35 C65 2022 | DDC 632/.9517—dc23

LC record available at https://lccn.loc.gov/2021047425

ISBNs: 9781645036746 (hardcover), 9781645036753 (ebook)

LSC-C

Printing 1, 2022

For my family

CONTENTS

Contents

PART III

INTRODUCTION

In the summer of 1944 Robert Mizell was leafing through *Time* magazine when an article in the Science section caught his eye: censorship had just been lifted on one of World War II's top-secret discoveries, a chemical called DDT.

At first glance, DDT sounded mundane. It killed bugs. Flies, bedbugs, moths, roaches, dog fleas, potato beetles, cabbage worms, fruit worms, corn borers, mosquitoes, their larvae, and more. But a US Army official said that by killing mosquitoes, DDT promised to wipe out malaria. He claimed it would revolutionize medicine as much as the discovery of antiseptics had revolutionized surgery. Mizell, a top administrator at a university in Atlanta, clipped the article and underlined the bit about malaria. He attached a note and sent it to an old college friend, Robert Woodruff, the soft-drink magnate who ran Coca-Cola.

"If the stuff is as good as reported, have you thought about the tremendous economic implications . . .?" wrote Mizell, who often advised his friend on business and charitable matters. "Maybe we should buy some cheap land. Say nothing about it." In the chemical he saw gold: ranches, housing developments, vacation resorts, and golf courses going in where mosquitoes and flies once thrived. Many like him did. A journalist for *Life* magazine reported that where US troops were stationed in the South Pacific, DDT had "proved that it could easily convert a verminous hellhole of an island into a health resort."

When the war was over, DDT came home a hero. It entered a booming postwar consumer marketplace, where it became the solution to a long list of postwar problems. Farmers sprayed it on orchards, vineyards, and croplands and dipped whole herds of cattle in it. Developers erected new suburbs using DDT-coated plywood. Home owners moved in and decorated with DDT-slicked wallpaper. They sprayed kitchens to kill ants and roaches, dusted mattresses to kill bedbugs, and treated pets to kill fleas. Dry cleaners added DDT to their cleaning solutions to ward off moths. Hotel and restaurant decorators arranged bouquets of DDT-impregnated fake flowers to repel wasps and bees. City officials sent out cavalcades of DDT spray trucks to clear neighborhood streets of insects, and children ran behind them, playing in the mist.

In just a few short years, the pesticide—a relatively simple compound of carbon, hydrogen, and chlorine, used with abandon—had come to symbolize our postwar nation's capacity to vanquish age-old scourges with modern science and technology.

Three decades later, it was banned.

•

The 1972 ban—technically a regulatory restriction of DDT's approved uses—followed years of mounting protest boosted by a nature writer named Rachel Carson. Her 1962 book *Silent Spring* implicated all of the new postwar pesticides, DDT included, in an epic attack on American wildlife. DDT was a neurotoxin with a predilection for fat tissue and a tendency to stick around, or persist, long after it had been sprayed. It killed beneficial bugs, fish, and birds. Carson also speculated that its broad class of chemicals, the synthetic postwar pesticides, was responsible for the nation's rising number of cancer cases. But she was mostly concerned with what DDT had come to symbolize to her: the nation's rush to embrace quick-fix technologies without taking the time to learn about their unintended consequences, especially those that weren't immediately apparent. Following high-profile hearings held by the nation's brand-new

Environmental Protection Agency, DDT left the market with as much fanfare as it had arrived.

Then, a generation later, a seemingly grassroots movement rose up to call for DDT's return. DDT's defenders argued that the chemical was the best tool against malaria, which was resurgent in sub-Saharan Africa. They also argued that its harms had been gravely overstated. Rachel Carson was wrong, they said. It was time to bring back DDT.

That didn't happen, but when more than a hundred nations signed an international treaty in 2001 to phase out another class of chemicals to which DDT belonged, the persistent organic (that is, carbon-based) pollutants, they did carve out an exception for the chemical. DDT, the signatories agreed, was critical for public health, even if it was known to be toxic.

•

That's DDT's story in three neat acts: war hero turned pariah turned exception. For a historian of medicine like me, it's a familiar story; it's one I've shared with students many times. But the third act always nagged at me. Why was the late 1990s the moment when Americans suddenly started calling for DDT's return? A few years ago I decided to try to figure that out. In a handful of emails and letters in a collection of corporate documents, I found an unexpected answer.

In the late 1990s, public relations specialists for Philip Morris—the tobacco company—were compiling a list of the century's most important and inspiring women as part of a stealth campaign to promote Virginia Slims cigarettes. Rachel Carson was on the list for her groundbreaking work exposing the dangers of pesticides such as DDT. At the same time, however, Philip Morris executives were funding an entirely separate campaign, one to bring back DDT.

The tobacco industry had no interest in selling the pesticide, of course; it was trying to sell cigarettes. But it found DDT's story to be a helpful scientific parable, one that, told just right, illustrated the problem of government regulation of private industry gone wrong. DDT, in this tale, never should have been banned in the first place. Companies, not liberal activists and politicians, should be trusted to make responsible choices. DDT's fate showed what happened when government got in the way. To sell one poison, in short, the tobacco industry sold a morality tale about another.

DDT's story was also, for the tobacco industry, useful for a far simpler reason: it was a distraction from the accumulating science on the dangers of secondhand smoke. Distraction is one of a list of tactics that various industry players have long used to protect markets for their products. Distract public attention away from unfavorable evidence. Discredit scientists and evidence you don't like. Distort findings so they say what you want them to say. Deny evidence that isn't in your favor. These strategies take advantage of the debate and uncertainty inherent to the scientific process. They also capitalize on the sensationalist tendencies of our news media and exploit the public's dependence on the media for its understanding of scientific issues. "Discredit a scientist," advised a set of guidelines drawn up by a Philip Morris front group in the nineties, "but don't spread the word yourself. Get the news media to do it."

This set of tactics has a much longer history than the last few decades. The tobacco industry first began sowing scientific doubt back in the 1950s to deny the then-emerging harms of smoking. It turned to public relations specialists who had just begun to develop strategies for the chemical industry, which was facing Congressional scrutiny. In the decades that followed, the two industries and their fellow free-market defenders—conservative think tanks and other sectors seeking to limit government regulation—doubled down and expanded on those mid-century PR efforts. In the process, they obliterated the US public's trust in science. They stoked today's climate-change doubts, GMO stalemates, vaccine fears, and COVID denial. They led us to the moment we're now living in, a moment in which science is intensely polemical and politicized.

·

This revised understanding of DDT's third act left me wondering about its first two. As I looked into them anew, I found material that complicated the stories of DDT's postwar popularity and 1972 ban.

And as I pieced it all together, the picture that emerged shed even more light on the contested nature of science today.

In Act 1, DDT's rise has long seemed to illustrate postwar Americans' faith and trust in science and scientific expertise. Countless scientists and citizens certainly embraced DDT during and after the war, but many did so because they had to, or because they believed they had to. Those who were openly critical of DDT, meanwhile, often found themselves dismissed as ignorant or even mentally unwell. And while DDT spelled profits for those with the purchasing power of a Woodruff, its rise was linked to larger economic shifts that stripped people living at the economic margins of their land and livelihood, in the process eroding their trust in the scientific and government experts who promoted DDT.

Act 2 has long been read as a moment that captures the ascent and power of environmentalism. But that account leaves out the economic forces at play behind the scenes. By the time environmentalists turned regulators' attention to DDT's downsides, the larger chemical companies wanted DDT off the market so they could sell pricier, patented pesticides. Tobacco companies wanted DDT out of farmers' hands because it was threatening US tobacco sales abroad, where other nations were already restricting the amount of DDT permitted in products. Scientific proof of environmental harm shifted policy only when other interests aligned.

Throughout all three acts, meanwhile, DDT's manufacture, use, and persistence contaminated soil, rivers, creeks, and oceans. The chemical and its breakdown products entered food webs and human bodies. Scientists labored to disentangle its effects from that of other chemicals and aspects of modern life. Scientific uncertainty, paradigm shifts, and plain old pride and ego complicated the task. Citizens struggled to shield themselves or their communities from DDT pollution. Their battles often ended up in the courts. All the while, DDT built up in soils, waters, and bodies in amounts determined by the social and political forces connecting place, class, and race.

DDT's three acts, reconsidered, struck me as a story about how science and its practitioners—who once not just promised but assured us that DDT was safe—can be comforting to some people and suspect to others, not because they're uneducated or uninformed, but because they see the world in a different way, and because they have every reason to. It's a story of how science becomes the turf on which we do battle over differences of gender, race, economic power, and more—without ever admitting as much. It's a story, all told, that shows why we fight about science—and why science has the power to divide us.

Prologue

FISH FOR THE TABLE

On the outside, Clyde Foster kept it cool behind crisp collars, a trim mustache, and close-cropped hair. On the inside, though, he often felt like a stick of "dynamite" waiting to explode.

Then one day he did. The spark was a story buried in that day's paper: "Danger Seen in Eating Fish from Rivers." A government survey of streams in seven southern states had found "huge" quantities of the banned pesticide DDT in a creek feeding the Tennessee River outside of Huntsville, Alabama. An estimated four thousand tons of the chemical lay settled at the bottom of a two-mile stretch of the creek known as the Huntsville Spring Branch. Fish in the creek, the survey found, carried enormous amounts of the pesticide in their bodies. The source of the contaminant was an old manufacturing plant at a nearby arsenal. The plant had been defunct for years, but it was still seeping chemical residues into the creek, which met up with the Tennessee at a bend in the river where the small town of Triana sat. Foster's town.

Triana was Foster's town in more ways than one. His wife, Dorothy, had grown up there, and her family told proud stories of how it had once been a bustling place, with a cotton-shipping port busier than Huntsville's, a hotel, and a saloon. But when Clyde and Dorothy had moved there in 1957 for his job at the nearby Army Ballistic Missile Agency, he was struck by the town's deprivation. At the space center, engineers, scientists, and analysts like him prepped rockets for Moon launches. Fifteen minutes down the road in Triana, however, people still lived without electricity or running water.

Foster, a soft-spoken man nonetheless known for his determination, decided to do something about it. He traveled to Montgomery to dig a copy of Triana's nineteenth-century charter out of the state archives. He found a judge to reinstate the charter so that he could apply for government grants to rebuild the town. The judge's order put Triana back on the map and made Foster, at age thirty-two, the town's first mayor since the nineteenth century.

By then, it was 1964. At work, Foster analyzed weather data at the missile agency, now part of the new National Aeronautics and Space Administration, for rocket launches. Nights and weekends, he brought Triana into the twentieth century, putting in streetlamps, hiring a police chief, and creating programs to give people a crack at the kind of secure, middle-class job he had himself. Gradually, the small town grew, reaching around a thousand residents by the late 1970s.

But it remained poor. And for most of the town's residents, poverty meant living off food from the land, including fish from the Tennessee River and its creek. Which is why, in 1978, the buried headline incensed Foster to the point where he lost his characteristic calm. "We've been eating the fish from that water for years and years, and we're just now learning that it has all this DDT," he said. "These are table fish, not trophy fish. These fish were caught to eat, not to show off."

Poisoned fish isn't what Foster expected would set him off. At the space center, being Black meant an endless assault of discrimination and injustice. Social functions for white colleagues were off-limits. So were trainings, and therefore promotions. He faced the same as mayor when local whites asked him if he intended Triana to become an "all-Negro town."

As he started digging into the story of how DDT got in the river, however, a legacy of unmistakable racism unfolded before him. He learned that the US Army, which owned the land containing the former DDT plant, had known about the contamination since at least 1964. The Federal Water Quality Administration had known since

1969. The Environmental Protection Agency had warned against eating fish from the river for a whole year. But no one had told anyone in Triana. "If this community had been anything other than Black," said Foster, "the circumstances would have been different."

But Foster had become mayor at the height of the civil rights movement. He had watched as the federal government stepped in to help desegregate Montgomery's buses and defend Birmingham's peaceful protesters, and then he had watched as President Lyndon Johnson signed the Civil Rights and Voting Rights Acts into law. He believed that someone in the federal government would make things right if they knew what was going on. And he had an in: he was a federal employee.

He picked up the phone and began making calls. Within a week, he was sitting in a meeting with scientists from two federal agencies, the Tennessee Valley Authority (TVA) and the Center for Disease Control (CDC). The TVA, created by President Franklin D. Roosevelt back in the thirties to spur economic development in the area, urged Alabama officials to ban any further fishing. The CDC agreed to start testing fish stored in residents' freezers to determine exactly how much DDT the people in Triana had already consumed.

The pesticide was, in chemical terms, a chlorinated hydrocarbon, a compound containing chlorine atoms attached to carbon rings trimmed with hydrogen. DDT's particular arrangement of atoms gave it staying power: when sprayed on crops or trees, it stuck around. It was also fat soluble, which meant that it lingered in the fat of living things. In food chains, this meant that as DDT "bioaccumulated" in individual organisms, it also biomagnified, building up even more in organisms higher up in the food chain as they consumed all the DDT eaten by organisms below them on the chain.

The Food and Drug Administration (FDA) had thus long before set limits, officially called tolerances, on the amount of DDT allowable in the US food supply. But when a family caught its dinner from a nearby river, those tolerances were all but meaningless. The CDC's initial tests revealed that residents were eating fish with DDT levels fifty times higher than what the FDA considered safe.

A second round of tests, by the TVA, found fish with ninety times the safe level.

Foster knew that the people of Triana ate hundreds of pounds of fish each week; Dorothy cooked more than six pounds a week for their family alone. With that much fish in their diet, he knew that they all had the chemical in their bodies. What he wanted to know next was how much—and what, exactly—it was doing to them.

The CDC was also curious. The agency's own scientists had vouched for DDT's safety back in the forties and fifties, but scientific thinking on chemical risks had changed since then. The agency also had amends to make: just a few years before, it had settled a lawsuit with participants in its long-standing "Tuskegee Study of Untreated Syphilis in the Negro Male," carried out long after the vicious disease was actually treatable. Now the agency quickly dispatched an epidemiologist from its Atlanta headquarters.

Kathleen Kreiss, a careful Radcliffe grad from Nebraska with glasses and a middle part, traveled to Triana to collect samples. She took blood from Foster's wife, Dorothy; his police chief, Joe Fletcher; a town elder named Felix Wynn; and nine others. Back in Atlanta, she found that all twelve residents had as much, or more, of the pesticide in their blood as did DDT factory workers, who, until Triana, were thought to have the highest DDT levels in the world. Worse, some of the residents had more—much more. Eighty-three-year-old Wynn had four times more DDT in his body than the highest level ever recorded in any human. And neither he nor any of the others had ever worked at the arsenal plant.

Wynn and the others received their results by mail, on official letterhead from the CDC. No studies, the letters assured them, had ever found any health problems in DDT factory workers. But Kreiss, troubled by the findings, traveled back to Triana to meet with the group in person. She wanted to personally reassure them that all Americans had some DDT in their body because the pesticide had been sprayed on so many farms, fields, and towns for so long. And because no studies had ever found any significant health problems from DDT in the US population—or even in highly exposed factory workers—it was very likely that the people of Triana were going to be just fine.

But the people were far from convinced. The news that you had enormous amounts of DDT in your body "did something to you mentally," said Fletcher. "It drained you down. You'd be out walking, and all of a sudden it would hit you that you had this stuff in your blood—kind of like cancer." Everyone knew that DDT was a banned chemical. Six years earlier, President Richard Nixon's newly formed Environmental Protection Agency (EPA) had pulled it off the market following highly publicized, years-long hearings. Announcing the ban, EPA head William Ruckelshaus had cited evidence that DDT harmed wildlife and carried a risk of cancer in people. News reports on the hearings had announced that DDT possibly caused other problems, too, including hormonal imbalances and birth defects. All of this made the CDC's reassurances hard, if not impossible, to trust.

Foster asked Kreiss for her support in asking the CDC to set up a full-time research center in Triana. In less time than it took to get his town's old charter reinstated, the agency agreed. From a scientific standpoint, the question of DDT's harms to humans was a still-unsettled one despite the studies of factory workers, and Triana offered a unique opportunity to understand DDT's risks. The town, a local magazine announced, was going to "make medical history by providing the first massive data on the effects of long-term exposure to DDT in human beings."

Those answers, however, would prove to be a long way off—longer, certainly, than anyone imagined at the time. In the meantime, Triana became a battleground for a scientific dispute over just how toxic the banned chemical actually was. Its noxious effects on wildlife were well-known. But Olin Chemical Corporation, the company that had acquired the company that had long run the plant on the arsenal, insisted that DDT was no more dangerous to people than table salt. World-famous toxicologists, experts in the harm that chemicals do to living things, agreed. Epidemiologists and risk scientists insisted that they were wrong, that the risks of long-term harm were potentially real. Medical tests on the people of Triana continued, and they came back inconclusive. Nobody, said Fletcher, seemed to know the truth about DDT.

In the meantime, Triana residents needed to figure out how to live with a painful paradox of American life. They had been living in nineteenth-century conditions while a twentieth-century chemical had made its way into their water, their food supply, and their bodies.

As they struggled to adapt to the contaminant within, DDT continued to leach from the site of the plant. Whole crystals were visible on the banks of the creek. DDT's slightly sweet smell permeated the air. It continued to accumulate in the sediment of the river and the bodies of fish and birds that seemed to somehow survive it. But their survival offered little comfort to Marvelene Freeman, who owned a local grocery store and whose five children all tested positive for high levels of DDT.

"This used to be a lively little town," she said. "But this problem has changed everyone's attitude. We're like prisoners on death row, just waiting to die."

PART I

Chapter 1

NOT TOO MUCH

Victor Froelicher had prospered during the Great Depression. The Swiss-born chemist and his wife, Helen, bought a gracious new slate-roofed home on a tree-lined street in Ridgewood, New Jersey, just outside of New York City. They raised five children on its half-acre lawn and hosted endless events within its plastered halls: holiday socials, buffet lunches, community meetings, and concerts. Victor played the organ and directed their church choir; Helen raised funds for charity. Their lives felt full and generous.

Froelicher owed his prosperity to the Textile Dyeing and Printing Company, which had a plant in nearby Fair Lawn, where he was chief chemist. Textiles were big business; even during the Depression their production and consumption rose. As the national economy recovered, though, struggles between management and unions rattled the company. In 1938 the Fair Lawn plant shuttered, and Froelicher was out of work. He traveled back to Switzerland, leaving his family behind, in hopes of finding new work to support them. Before a year had passed, he was back in the states with a new job: he was the US representative for the reputable Swiss chemical company J.R. Geigy, which had been looking to expand its reach into the US market for decades.

Geigy was one of the oldest and most successful of the big dye companies in Europe, founded all the way back in 1758, when it began as a manufacturer and trader of not just dyes but also spices, drugs, and other chemicals. A century later, the company joined a pack of German firms that began making dyes and drugs from coal tar, an abundant waste product of the Industrial Revolution. It was a booming business. By the late 1800s, Geigy had sales offices in Europe, Asia, and North America. In the early 1900s, Geigy began setting up plants in the United States. By the late 1930s, just as the Textile Dyeing and Printing Company was shutting down in Fair Lawn, Geigy was closing its own plant in Jersey City but only to build a much bigger plant in Bayonne. The company also had plans to erect a research lab next door to the new Bayonne plant.

This new plant, Geigy planned, would focus on the next big thing in chemistry: compounds that killed insect pests. Insects had assumed pest-scale proportions for more than half a century by then. Massive changes in US farming and transportation in the late nineteenth century had invited a long list of destructive bugs: the Colorado potato beetle, boll weevil, gypsy moth, codling moth, cotton army worm, plum curculio, and countless others. White cabbage butterflies destroyed cabbage crops. Joint worms wiped out wheat crops. Rocky Mountain locusts devastated just about everything in their paths, sometimes swarming so thick that they slowed trains on newly laid tracks. Farmers picked insects off plants by hand, shook them off with wheeled carts, and crushed them with blocks and bricks. They loosed natural parasites and predators of the pests, sometimes creating bigger problems in the process. Desperate, they also applied poisonous chemicals, especially lead arsenate and the copper-arsenic compound known as Paris Green.

The poisons were effective. Straight arsenic killed the cereal-destroying army worm. A sprinkling of Paris Green could kill half the locusts in a field. They had just one problem: their residues clung to crops. And on crops eaten fresh—like fruits and vegetables—such residues could sicken and kill.

By the 1920s, unwitting consumers began to fall ill, left shaking with fever, bloody urine, and "thready pulse" after eating asparagus, cabbage, celery, or the like. Some, like the Montana girl who ate three apples in a row, ended up dead. Journalists reported that people were "refusing to touch green commodities." Doctors warned of a "menace to the public health." By the time Froelicher stepped into his new role at Geigy, there was a frenzied search for a generation of insecticides that killed pests but not people.

Geigy's Bayonne lab was still relatively new when World War II began, but its mandate was already clear. Insects weren't just a threat to agriculture; they were a menace in war, too. They feasted on the destruction left by combat and were drawn to hordes of soldiers and refugees in crowded, unsanitary conditions. Lice spread the aches, nausea, and delirium of typhus. Mosquitoes spread the bone-rattling fevers of malaria along with dengue, yellow fever, and filariasis. And the US knew this from experience: when the country entered World War I, malaria-carrying mosquitoes had ravaged troops sent to train in the US South, hobbling forces before they even shipped out. In the Civil War, insect-borne and other infectious diseases had killed more soldiers than combat. Everyone knew, as one observer put it, that the American and British governments "were desperately looking for a means of keeping insects away from their troops."

•

On an early fall day in 1941—with Europe in tatters and German forces moving across the Soviet Union—Froelicher arrived at work to find a report on his desk from Geigy headquarters in Basel. A chemist there had discovered a dust that was "extremely effective" against the Colorado potato beetle. The dust had saved the Swiss potato crop, reportedly forestalling a famine in Switzerland. Froelicher still had relatives there, so the news was a source of personal relief. Professionally, though, it "did not seem to be extraordinarily important," he thought. The US potato crop was doing fine. Lead arsenate kept the beetle in check when it did appear. Washing and cooking helped managed the residue problem. And although wartime shortages had left lead arsenate in short supply in Switzerland, in the US, which had not yet entered the war, the insecticide was abundant.

Two months later, the Japanese bombed Pearl Harbor. Less than a year after the United States went to war with Japan in the Pacific and Germany in Europe, Froelicher arrived at work to find

17

yet another report on the chemical. This time, he paused. The insecticide was now killing "a great variety" of crop pests. It had spared Switzerland's war refugees from infestations of disease-carrying lice. Intriguingly, it worked two ways: bugs died when they ate it, but also when they just touched it. Froelicher knew that the US Department of Agriculture was very interested in finding the holy grail of insect killers, one that was both a "contact" and "stomach" poison. The agency had tested thousands of contenders without success. The possibility that this new chemical might be it, he thought, "seemed altogether too good to be true."

But worth exploring. Malaria was spreading faster among troops in the Pacific than it ever had in the US South. Troops and refugees in the Mediterranean were being ravaged by typhus. Froelicher requested a sample. Headquarters sent him a little over a hundred pounds of the chemical in dust form, which they called Neocid, and the same amount of a spray form, dubbed Gesarol. The samples arrived in New Jersey plainly wrapped, two anonymous packages bearing no details about their ingredients or composition. As he pondered what to do with them, another communication landed on his desk. A US military attaché in Berne had heard about the dust's capacity to kill lice and thought that Allied forces should know about it. Froelicher quickly devised a strategy to test and distribute the chemical. He kept some for himself and forwarded some to the USDA's Division of Insecticide Investigations to see if it could verify Basel's claims.

The USDA division had a team of entomologists that had been hastily assembled after the bombing of Pearl Harbor and stationed in muggy Orlando, Florida, where lice, ticks, flies, and mosquitoes thrived. Working out of a shed of a building near Orlando's T.G. Lee Dairy, the entomologists had developed a handful of repellants and insecticides. None of them were very remarkable. The scientists' biggest accomplishment to date had been to create massive colonies of mosquitoes and lice. But the lab had a relatively new director, Edward Knipling, a thirty-five-year-old Texan with five kids, experience controlling bollworms in Mexico, and a reputation for being patient, thorough, and precise. At any given time, Knipling's four dozen scientists were testing more than a hundred insecticides on their colonies. They had a routine protocol for evaluating new insect killers, and Froelicher knew they would run the Geigy compound through it.

The protocol started with a sleeve. When Froelicher's samples arrived, one of Knipling's scientists coated a sleeve with the Neocid dust, filled it with some of the lab's lice, slipped it on his arm, and went about his day. When he removed the sleeve the next morning, the lice were dead. He filled the sleeve with fresh lice and put it back on for another day, and the next morning the new lice were dead, too. Some of Knipling's other scientists took the Gesarol into town, where they sprayed it on a theater with a long-standing cockroach problem. The roaches disappeared. They sprayed the mess halls at a nearby army base next, eliminating its flies. Other scientists on the team went over to the nearby citrus groves to recruit pickers willing to let the lab's lice feed on them. They dusted half the volunteers with the Neocid first, and their lice died overnight.

It wasn't that unusual to find a poison able to kill both cockroaches and lice—after all, arsenic could. But the theater didn't need to be resprayed the next day. Or week. Neither did the mess halls. Six weeks later, in fact, the researcher with the sleeve was still finding his fresh lice dead by morning, and he hadn't applied any more of the dust. Then the insect killer really surprised the team. They put it on wool underwear and gave the underwear to volunteers heavily infested with lice. Then they washed the underwear and gave it back to the volunteers—and the *washed* clothes kept killing new lice. At that point, Knipling felt a small rush. "Perhaps then we were a little excited," he said.

The team still had no idea what, exactly, they were working with, so Knipling asked his head chemist to extract the active ingredient and send it to a federal lab in Beltsville, Maryland, for analysis. The Beltsville chemists sent back the answer: 2,2-bis (parachlorophenyl) 1,1,1-trichloroethane, more easily described as dichlorodiphenyltricholoroethane. Knipling shared the chemical structure with Froelicher, who, with his doctorate in chemistry, found it baffling. The compound's assemblage of carbon, hydrogen, and chlorine atoms "suggested no insecticidal properties at all," he said. "Cap this with the fact that the discovery had been made by J.R. Geigy, Switzerland, well-known pioneers in the synthetic dyestuff field, but relatively unknown in the insecticide field." He ferreted some of the "organic"—that is, carbon-based—dust from his office to his home in Ridgewood and prepared for a hard sell.

Back in Orlando, Knipling's team kept testing, sworn to secrecy as they did. They constructed tiny bedrooms with wood walls and miniature mattresses, dusted them with the chemical, introduced bedbugs, and watched the bugs die. They treated full-sized mattresses in Florida army barracks, giving the soldiers there "complete freedom from bedbugs" for months. They dusted forest lots jumping with chiggers, clearing them of the biting mites. They fed the chemical to a rabbit, shaved its abdomen, let ticks feed, and watched the ticks perish as they consumed the chemical in the rabbit's blood. They dusted a cocker spaniel infested with thousands of ticks; his itching eased. They applied a paste made from the poison to cattle's ears; their ticks succumbed. They sprayed the chemical in nearby barns and Gulf Coast horse stables, putting a previously unfathomable end to their notorious infestations of flies.

The team's tests, said Knipling, "fully confirmed" Geigy's claims about the compound. Moreover, the chemical didn't appear to be poisonous to any of the animals they tested it on. He just wanted to be sure it was safe for people, too.

But the height of war was no time to wait for more evidence. General Douglas MacArthur, who was commanding forces in the Pacific, was said to be more worried about mosquitoes than the Japanese. Knipling shared his team's results with Froelicher and the Department of Agriculture. A USDA official shared the findings with the US surgeon general, Thomas Parran. A powerful, outspoken, and demanding administrator, Parran had had a hand in drafting the Social Security Act and had launched a national crusade against syphilis, a disease previously managed with a combination of moral condemnation and toxic, arsenic-derived drugs. Parran made it well-known that he had but one duty in his post: to save as many American lives as possible.

Armed with Knipling's test results, Froelicher headed to Washington to meet with Parran, packing with him an unmarked sample of pure dichlorodiphenyltricholoroethane. Froelicher knew that his task was to convince Parran, over anyone, that the substance could prevent disease without causing harm. He described the Orlando team's findings in detail and spoke of the relief the chemical could grant soldiers from discomfort and dreaded diseases. But he also knew that to truly convince Parran, he would need to prove the compound safe. His knowledge of chemistry, combined with Knipling's tests, emboldened him to put on a bit of a show. Seated across from Parran, Froelicher took out his sample, showed the powdery crystal to the general, and then ate a chunk the size of a walnut. "Dr. Froelicher," said Parran, "if you are alive tomorrow, you will be getting a big order from us."

•

Froelicher was, indeed, alive and well the next day. Very quickly, he got word from Washington: the US Army was "very interested" in the compound. Froelicher wasn't surprised. The chemical, he knew, had the potential to alter the course of the war.

Geigy hadn't shared any manufacturing instructions, but with the chemical analysis from Beltsville they had been relatively easy to figure out. Knipling wrote them up and published them in the scientific literature: chloral hydrate or chloral (a well-known sedative) needed to react with monochlorobenzene (a common industrial solvent) in the presence of sulfuric acid (a chemical widely used in synthetic dye manufacture). Geigy immediately began preparing for pilot production at a pair of factories on the Ohio River known as the Cincinnati Chemical Works, which Geigy had set up with Swiss partners Ciba and Sandoz back in the twenties to avoid US tariffs on European imports.

But so much was still unknown about the chemical. For one, no one knew how it worked. Back in Basel, Geigy scientists closely tracked what happened to insects when they encountered it. When flies landed on a lightly sprayed surface, the chemical entered their bodies through delicate receptors on the tips of their legs. They became highly excited and then slowly paralyzed, until they turned on their backs, legs feebly waving in the air, and died. A May beetle, by contrast, could walk on DDT unharmed, but a mouthful of DDT-dusted leaf paralyzed its jaws and digestive system. The beetles died of starvation. Butterflies that landed on DDT sometimes abandoned their legs, which would tremor and convulse violently on their own for hours. The compound was an indisputable nerve toxin.

Based in neutral Switzerland, Geigy hadn't reserved DDT for the United States. The company had shared it with the British and with the Germans, too. British scientists found it impressive, and it was a British official, typing in breathless haste, who shortened its cumbersome chemical name to DDT. German scientists were just as impressed, but they held off on any further research, worried about the compound's potential risks to people.

The question bothered Knipling, too. It was, he said, "our chief worry." So his lab partnered with researchers at the Rockefeller Foundation in New York to test the compound on larger populations than his team had access to. The Rockefeller scientists dusted and sprayed DDT on medical students in New York, conscientious objectors in New Hampshire, and civilians in Mexico. Not a one so much as sneezed or itched. But it would take a much bigger population for Knipling to feel confident that DDT was safe.

In December 1943—just over a year after the first package of DDT had arrived in New Jersey—Knipling found a larger population on which to test the pesticide. In October, Allied forces had defeated German troops in Naples, Italy, and the fighting had left the city in shambles, hundreds of thousands of residents and refugees on the streets and in camps with no water, gas, or electricity. The damp, filthy, crowded conditions had triggered an explosion of lice. A typhus epidemic, army medics worried, was imminent. Knipling heard about the situation, and he boarded a plane for Europe. Froelicher heard as well, and Geigy's Ohio facility ramped up production.

In Naples the Rockefeller Foundation and the army set up forty-two delousing stations across the city, each equipped with a set of dusting guns that released a puff of DDT with each pull of a trigger. It took two minutes to fully dust a man: a soldier inserted the tip of his gun into each sleeve, down the neck of his shirt, inside the front waistband of his pants and then the back, and in his hat or coat if he had one. On a single day, the stations dusted 72,000 civilians. By the end of January, they had dusted 1.3 million. The feared typhus epidemic never materialized. And the dust hadn't harmed a single person. DDT, announced the British, was going to give the Allies "a higher degree of security against the dangers and discomforts of insects than any army has ever had."

After the Naples experiment, Knipling's mind settled. Back in Orlando, his team set up an "insecticide school" and outlined a new protocol, this one to guide army personnel on DDT's uses. It started with instructions on how to make a spray: by dissolving a little bit of pure, or "technical," DDT in kerosene, fuel oil, diesel oil, or waste crankcase oil from vehicles and planes. It was important to keep the DDT concentration low, they noted, because too much could kill fish in nearby waters. (This bothered no one; after all, lead arsenate and Paris Green were capable of the same.) The next step was to spray a one-mile radius around each cantonment area. After that, dust mattresses. In combat, toss DDT-containing aerosol bombs into foxholes and gun emplacements. Afterward, spray battlefields full of corpses. As needed, dust soldiers, prisoners, refugees, and civilians infected with scabies, head lice, or pubic and crab lice. Tell them to keep on their clothing for several days so that DDT could permeate it. Tell men with head lice to leave the dust in their hair for up to three weeks.

Up in Washington, DC, meanwhile, Food and Drug Administration chemist Herbert Orion Calvery had another team of scientists assessing DDT. A former physiological chemistry professor at the University of Michigan, Calvery spent the war years working for the government. As chief of the FDA's Division of Pharmacology, he had his own approach to testing new drugs, food chemicals, or insecticides for safety: it involved applying and feeding incrementally larger doses to a range of lab animals to determine the doses at which the chemical became toxic, and then deadly.

What Calvery's scientists saw as they fed DDT to a long list of animals—including guinea pigs, pigeons, rabbits, and dogs— troubled them. DDT dust didn't absorb through the animals' skin, they noticed, but DDT in spray form soaked right in. When it did, it poisoned some of the animals but not in obviously predictable ways. The amount that triggered tremors in a cat, for instance, might do nothing to a similarly sized dog. They also noticed that if they fed small animals repeated small doses—small enough to be nontoxic—the animals eventually experienced toxic effects anyway. In that regard, DDT was reminiscent of lead, which, physicians and chemists like Calvery knew, the body stored over time.

The FDA team's observations in Washington led Calvery to a very different conclusion from those that had come out of the USDA in Orlando and the army in Naples. DDT is "a poisonous substance," he determined. "It should be handled with care and its use should be carefully restricted."

Calvery sent his warning to US Army officials, who appended it to the end of a thirty-page restricted memo on DDT that the army released shortly after the Naples experiment. Along with his warning, Calvery asked for thirty-five weeks to test the chemical further. The officials agreed—in part. Calvery could keep testing DDT, but as he did so, the army was going to go ahead and use it.

Over the next several months, Calvery's FDA scientists found that large doses of full-strength DDT caused nervousness, convulsions, and death when fed to guinea pigs, rabbits, and other lab animals. Dissolved in solvent and applied to rabbit bellies, it led to tremors and paralysis. Feeding it or applying it as an ointment to various lab animals damaged their kidneys, livers, and testes, and led to gastric bleeding. The effects still varied by species; inhaled DDT killed mice, for instance, but left monkeys and dogs unharmed. But across all species they studied, one thing seemed consistent: the animals stored up DDT in their fat. The pattern explained DDT's ability to slowly poison an animal over time.

Calvery found the manner in which DDT collected in fat most "alarming," but he was also conflicted. He knew, of course, that all chemicals were toxic at some dose and that wars justified risks that wouldn't be tolerated in times of peace. DDT was clearly preventing disease; and yet he believed that its use was also "fraught with hazard."

•

Down in Orlando, at the army air base near Knipling's lab, a twenty-eight-year-old white man from New Jersey, who had been inducted at Fort Dix the spring before, was handed a duster gun, canister, nozzle, and instructions. Following orders, he sprayed rooms across the base with a DDT mist so thick he couldn't see six feet in front of him. His clothes were soaked minutes into each day. A white film clung to his pants and shoes as they dried. He dusted over and under every mattress in the barracks. His eyes teared. His nose and throat scratched. At night he showered, and in the morning he pulled on his fatigues from the day before. Three times, army medical officers came to take samples of his urine and blood. One time, a doctor for the DuPont company came and listened to his heart.

He didn't know it, but the army had identified him as one of three known men with the greatest exposure to DDT. He was one of a handful of living experiments through which the army sought to resolve the contradiction between the USDA's and the FDA's conclusions about the chemical.

DuPont was grateful to have access to him, for the federal War Production Board had just asked the Delaware chemical company to start making DDT. The board was tasked with protecting the economy so that production of commodities needed for the war effort wouldn't flag or falter. It pressed private companies into producing materials needed for the armed forces, from aircrafts and ships to penicillin and nylon. It also set policies restricting civilian access to those same products and their raw materials. In January 1944, DDT was placed under the board's jurisdiction. That spring, along with DuPont, the board asked Merck and the Hercules Powder Company to start manufacturing DDT, too.

By that time, the medical officers' tests found that the Orlando army airman's heart, liver, kidneys, and skin were in fine shape. His psychological performance was average. He did better on steadiness tests than men with no exposure to DDT. The findings, their confidential report declared, "fail to indicate any definite evidence of toxic effects" caused by DDT. Army demand climbed. The War Production Board created a DDT advisory committee and pressed additional companies to begin making the chemical.

•

Back at the Froelicher home in Ridgewood, Helen and Victor awaited word from their oldest son, Charles, serving with the US 6th Armored Division in France. As they waited, Helen served in a Manhattan soup kitchen and helped German and Austrian Jews escape to the United States. Thousands of refugees passed through the living room of the slate-roofed house in Ridgewood, where the Froelichers served them beer and gave them loans in the thousands of dollars. When two young girls arrived, orphaned and displaced by the war, Victor and Helen adopted them.

Victor, meanwhile, ferreted more DDT samples back to the house. He dusted its comfortable rooms liberally. He placed whole crystals on display on the shelves of his study. His youngest son, Franz, all of six, was impressed with his father's story about eating DDT before the surgeon general. When his father wasn't looking, he helped himself to small pieces to snack on. Sometimes he did it for his own private thrill. Sometimes he did it to impress his play-mates. Most times, guilt would ultimately move him to confess the deed to his dad.

"OK," said Froelicher, unruffled. Just "don't do too much."

•

That June, in the summer of 1944, the US Marines invaded the Pacific island of Saipan, where insects swarmed so thick that it was difficult to see. The army's malaria-control units followed, covering ground torn up by battle to eliminate the stagnant, standing water that mosquitoes laid their eggs on. The men in the unit graded the earth with bulldozers, filled water-holding depressions with dirt, used draglines to eliminate swamps, and pulled choking weeds from streams so they could flow again. Rain came down in torrents. Wild boars and crocodiles threatened. Insects buzzed, bit, and stung. When long-awaited DDT supplies arrived, the unit soldiers dissolved the DDT in oil, loaded it into containers they wore like knapsacks, and sprayed any pool or puddle of standing water they could find, down to the circumference of a watch face.

It wasn't enough.

In August a dengue epidemic hit. For every battle injury, the station hospitals were admitting five cases of disease. The army consulted Knipling's team for advice, and they responded with a suggestion: spray DDT from the air. The men in the malaria unit loaded DDT emulsion into chemical-warfare tanks fitted to the bomb racks of two fighter planes. A crew flew them low over the island's palm trees. The tanks released a thin cloud of DDT that settled on every frond, leaf, thatch, and inch of ground. When they were done, Parran's assistant surgeon general announced that mosquitoes on Saipan were as rare as four-leaf clovers.

It was the beginning of DDT's victory in the Pacific. In the wake of each assault, malaria units sprayed dead bodies, food waste, latrines, and open pools of water. Trucks mounted with power sprayers rolled through next, spraying nets, hammocks, and camps. Planes followed, releasing a DDT drizzle from above. Malaria, dengue, and other diseases stayed at bay. The plans to invade Okinawa started with an aerial spray of DDT in diesel oil over 100,000 acres. On D-plus four in Iwo Jima, bombers released a coal-black spray of DDT in fuel oil overhead. Backup planes came up from Saipan and sprayed even more. Back home, headlines announced that US forces were "bombing" islands with DDT. Newspapers called the chemical a "miracle" bug killer, greater than penicillin, better than blood plasma, the "modern equivalent of the findings of Pasteur."

Halfway around the world, Charles Froelicher's division, dubbed the Super Sixth, advanced through France, liberating town after town after taking the beach in Normandy. By the summer of 1945, the Super Sixth had liberated the Buchenwald concentration camp in Germany. Charles, by then a master sergeant at age twenty-two, was handed a unique assignment. Given the possibility of finding relatives of his own, he was assigned to accompany a trainload of orphaned children from Buchenwald to Switzerland. Before his departure, he took a handful of photos of the shelled-out camp. He captured ashes spilling from a decommissioned oven, a heap of corpses on the ground, and a tangle of frail and expired bodies loaded onto a truck.

"Buchenwald 1945," he wrote on the back of the last photo. "This was Nazi Germany!"

•

The army's demand for DDT had mounted so quickly that by 1945, the War Production Board had pressed more than a dozen US chemical companies into production, from the now-forgotten Penn Salt Company to still-familiar Monsanto and Sherwin-Williams. Together, they churned out more than three million pounds a month. All of it, by the board's order, had gone to the US Armed Forces and the US Public Health Service. With the war's end in Europe and its end imminent in the Pacific, however, the board's DDT advisory committee pressed for coming DDT surpluses to be freed up for sale to the public. Its members represented the companies making DDT.

Government scientists pushed back. American and British army and navy personnel were sending back reports from the South Pacific of dead fish, crabs, prawns, dragonflies, and caterpillars. These echoed reports of dead tadpoles, butterflies, and birds from agency scientists studying the chemical in field tests in Georgia and Maryland. Department of Agriculture entomologists not connected with the Orlando field station worried that wider DDT use would harm bees and "the biological complex." Calvery warned that DDT might prove more harmful to humans in ways that simply hadn't become apparent yet. His scientists had just learned that when dogs ate small amounts of DDT regularly, not only did the chemical build up in their fat; it then also appeared, highly concentrated, in the milk they fed to their pups.

Calvery might have said more, but he fell so gravely ill that summer that he was forced to give up his position. Three months later he was dead, at forty-seven.

In August 1945 the board took the committee's advice and released DDT for sale to the public. The board wasn't ignoring scientific warnings; it simply had no justification for withholding DDT. The federal government didn't have the authority to keep an insecticide off the market; it never had. The most it could do was require manufacturers to honestly label any insecticidal products they sold—and issue a bland press release. The release made for tepid news of DDT's free-market debut. The front page of the *Washington*

Post stated that the War Production Board "warned against use of it to 'upset the balance of nature.'" "Careful Use Is Urged" ran the buried headline in the *New York Times*.

Not all of the wartime manufacturers thought the same way about commercializing DDT. Some, like Dow, were eager to capitalize on its wartime reputation. Others, like Monsanto, were somewhat hesitant, aware of DDT's risks to nature and worried that other downsides might come to light. Meanwhile, hordes of small manufacturers were already rushing to market. Some had started making and selling DDT before the board's release. It wasn't hard to do, noted a Georgia man with a chemistry degree and a fledgling bleach company. His college-age daughter helped him find the formula in a journal at the library. He used it to make DDT in his basement. Any "competent chemist" could do it, said a Swarthmore resident who sold his homemade DDT at local hardware stores. Local shipments to stores sold out before they even arrived.

City and state officials had been used to regarding DDT as a powerful chemical closely guarded by the US Army and government scientists. Now it was poised to be everywhere at once, and officials scrambled to respond. The Public Welfare Department in St. Louis issued a formal warning, calling DDT a "poison" whose effects on "plants, animals, and humans" were not fully known. Pennsylvania announced that because DDT was locally registered for only military use, it wouldn't be sold to the public at all. After all, said a state official, "DDT in any form is toxic to man." Minnesota withdrew DDT from the market. New Jersey allowed its sale only through pharmacists. New York and California ruled that any product containing DDT had to bear the skull-and-crossbones symbol, signifying a substance that could kill.

Before the summer's end, though, this patchwork regulatory wall crumbled. Pennsylvania—home to several wartime DDT manufacturers—began registering DDT for commercial sale. Minnesota began approving DDT-containing products one at a time. The changes came after the War Production Board announced that all three million pounds of DDT produced monthly by the wartime manufacturers would soon be available for sale. "DDT," a gleeful board representative told the press, "will be coming out of people's ears!"

Before DDT went to market, federal researchers and scientists tested it extensively, all at no cost to manufacturers. Now those manufacturers began to profit—even as federal research continued. Just not in Orlando. Four months after the war's end, Knipling's budget was cut 60 percent. His lab's entire wartime experimental budget was scheduled to end in less than a year. It wasn't personal; it was just that the war was over. Knipling himself was transferred to Washington and named chief of the USDA's Insects Affecting Man and Animals Division. He eventually picked up an old obsession, something he had been working on before the war, trying to see if he could eliminate a pernicious livestock pest, the screwworm fly, by sterilizing the males instead of spraying poisons. DDT had saved millions of lives in the war, he believed, "but there were just too many problems" with it.

•

Up in Ridgewood, Victor and Helen Froelicher welcomed Charles home. They made plans for his wedding to the daughter of the Italian consul-general of Philadelphia. The newlyweds honeymooned in Switzerland and Italy. Meanwhile, Froelicher's company made plans for another new plant, this one devoted solely to DDT. The plant sat on a 1,500-acre site in McIntosh, Alabama, adjacent to the Tombigbee River, a convenient place to release DDT by-products and other manufacturing waste.

The year after the war, Geigy and other DDT makers produced 45 million pounds of DDT. Overnight, DDT became the postwar answer to a long list of problems, from cotton weevils and tree beetles to ants in the sugar and fleas on the dog.

But perhaps the most important pest it attacked was one rattling the optimism of postwar American life, one thought to be transmitting a disease that spread like fire, devastating families, shuttering towns, and causing neighbor to recoil from neighbor. In desperation, those towns would drench themselves with DDT.

Chapter 2

POLIO CITY

In early June 1944, a Catawba County nurse transported a limp and feverish little girl from Hickory, North Carolina, to Charlotte's Memorial Hospital, an hour away. The girl's limbs had gone flaccid, a telltale sign of infantile paralysis, also known as polio.

Her case was just the first. Over the next two weeks the Charlotte hospital admitted more than twenty polio patients. Within a month, there were more than two hundred. Health officers declared an epidemic. They banned children from public and shut down schools, camps, pools, parks, and theaters. The faithful stayed home from church. People stopped shaking hands. Drivers kept their windows closed despite the heat.

Polio was a terrorizing disease, an infectious killer on the rise at a time when other contagious diseases were in decline. For decades polio had appeared in sporadic warm-weather outbreaks, but in the 1940s, as the US fought a global war abroad, it fought a war on polio at home. The disease returned one summer after the next, its case counts steadily rising. It struck mostly children but could paralyze adults, too. The worst cases immobilized patients from the neck down, consigning them to a full-body contraption known as an iron lung.

Polio had overwhelmed Charlotte Memorial Hospital twice before, in 1935 and 1942. Now it quickly filled to capacity a third time. The hospital erected one army tent after another to treat the

31

overflow. When those filled, Clarence H. Crabtree, state field director for the nonprofit National Foundation for Infantile Paralysis (NFIP), stepped in with plans to build an emergency hospital for polio alone.

Crabtree was a former state legislator with the calm, hooded gaze of a young George Washington, but his power didn't stem from his experience or looks. His employer, the National Foundation, widely known as the March of Dimes, was a $20 million organization with a massive public relations budget and widespread renown. Founded in the 1930s by Franklin D. Roosevelt, who had survived polio as an adult, it was headquartered in New York, with thousands of local chapters across the country. North Carolina had nearly a hundred chapters, and Crabtree oversaw them all. A Hickory native, he convinced town leaders to empty a summer camp on the outskirts of town. The State Guard helped prepare the grounds. The Red Cross sent nurses by the busload. Polio doctors and therapists came from New York, Chicago, and Philadelphia.

In three short, frenzied days, Crabtree's Hickory hospital was up and running. It admitted patients of any color, giving each one an army bed in an army tent—with army-issue screens to keep out the flies.

For although no one knew exactly how the crippling infection spread, flies had long been part of a popular theory. In the early twentieth century, when summertime outbreaks began ravaging cities, polio had acquired a reputation as a disease of filth. In crowded places such as New York, which suffered nine thousand cases and more than two thousand deaths in the summer of 1916, flies seemed to explain the disease's spread from "dirty" environments to clean ones. They also seemed to explain polio's preference for summertime, when flies thrived.

No one picked up on the irony that, as a historian later put it, "flies supposedly spread the polio germ by flying in only one direction, from the slums to the suburbs." At the time, plenty of scientific evidence supported the theory's basic premise. Harvard scientists had shown that flies could spread polio from one rhesus monkey to another. Other scientists had isolated polio virus from privies, sewage systems, and flies themselves. Still others had shown that feeding flies from epidemic areas to monkeys gave them polio. It wasn't clear how flies gave the disease to people, but Hickory presented an opportunity to figure that out.

As Hickory's new polio hospital filled with patients, the National Foundation's research department reached out to virologist John Rodman Paul, an exacting and fair-minded man who ran one of the top sites for polio research in the country, the Yale Poliomyelitis Study Unit at Yale's School of Medicine. The foundation offered Paul a grant to study polio's transmission in Hickory. Settling the question of how the disease spread was one of the foundation's top research priorities. Paul accepted the offer and dispatched three of his top scientists to the town to see what they could learn.

When Spokane-born Dorothy Horstmann boarded a train for Hickory, she was two years into a research fellowship with Paul at Yale. A thirty-two-year-old physician, she had just spent a year analyzing patient samples from a polio epidemic that had hit New Haven in 1943, to try to determine if and when poliovirus appeared in the bloodstream. Compared with New Haven, Hickory was hot when she arrived; it was also, she noted, poor. Outdoor privies were the "rule," and flies were "abundant." And the epidemic was quickly turning into one of the worst in recent memory, its hardships compounded by the tensions and shortages of war. Hickory's hospital was quickly stretched so thin that it soon admitted only the most serious cases, sending the rest to recuperate at home.

In a plaid dress and a chignon, a cluster of barefoot children trailing behind her, Horstmann went methodically through the foothill town, drawing blood from polio patients and setting flytraps outside their homes. The traps were baited with piles of sweetened banana mash. The idea behind them wasn't to catch flies but to collect the vomit and feces that flies left behind as they fed. Once the mash was contaminated, but before it spoiled in the heat, Horstmann collected it from the traps and shipped it back to New Haven, where it was fed to a pair of chimpanzees in Paul's lab.

Soon after eating the mash, the chimps—which the Yale team had named Catawba and Hickory—began passing poliovirus in their stool. They shed the virus for weeks. Back in New Haven that fall, Horstmann and her colleagues autopsied the pair. The primates' brains and spinal cords bore lesions typical of polio infection. The fly-contaminated mash, it seemed, had made them sick. It wasn't proof that flies spread polio to people, but it was evidence that pointed in that direction. The researchers summarized the findings and sent them to one of the nation's top research journals, *Science*, which published them in May.

"Poliomyelitis Virus in Fly-Contaminated Food Collected at an Epidemic," they titled the report. In the Piedmont town of Ruther-fordton, not too far from Hickory, a pastor named John H. Melzer came across the report in a college library. Like others in the region, Melzer dreaded polio's sure return. As he read about the chimps' disease, he put the findings together with the news from the war, with its images of bombers clearing islands of disease-carrying mosquitoes. If DDT was so good at killing insects, he wondered, why not use it to fight polio, too?

Melzer described the idea in a letter to the editors of the *Charlotte Observer*. Local papers across the Piedmont picked it up. In Hickory the *Daily Record*'s editors declared it an "excellent suggestion." DDT, they urged, should be used to combat polio "this coming summer."

•

The idea made perfect sense, not just because of DDT's role in the war but also because DDT was already being used by government officials to prevent disease across the US South. It had been ever since the Naples experiment. In fact, as Hickory's epidemic waned over the winter and its emergency hospital emptied of its last patients that spring, spray teams set out in thirteen North Carolina counties. The crews knocked on doors and asked home owners to put away food, take down mirrors and pictures, and pull furniture away from the walls so that they could coat everything with DDT.

Their target was a class of insects that spread disease—not polio, but malaria.

The antimalaria effort, a project of the new Malaria Control in War Areas agency, had nothing to do with polio at all. The MCWA was created by the federal government in 1942 to eliminate malaria from the grounds of military sites throughout the South, where so many recruits were being sent to train for war. Malaria, caused by a parasite transmitted by mosquitoes, was once found in every state east of the Rockies, but during the late nineteenth century and early twentieth century it had retreated to a swath of land that ran from East Texas to the Carolina coast. By the 1930s, it persisted largely in cotton-growing areas, where it was most common among Blacks living in impoverished conditions in poor housing near swampy land. Malaria rates soon tumbled there too, losing purchase as Blacks migrated to northern cities, window screens became common, and federal programs eliminated breeding grounds and moved people out of malarial areas. In 1943 the US saw just eight hundred malaria deaths.

But as the war in the Pacific ended, MCWA's scientists predicted a malaria surge in the United States. The agency had kept domestic malaria in check during the war by draining and filling lakes and ditches, and by coating fields and swamps with diesel oil and Paris Green, to kill mosquito larvae and adults. Yet two million Americans, they estimated, would soon return from war with malaria parasites in their blood, which local mosquitoes would pick up and spread. Malaria, they said, was about to ratchet back up to historical levels.

When MCWA scientists learned of DDT, it suggested a quicker and easier means of malaria control than earth moving and Paris Green. But the army gave them only limited supplies. So instead of spraying it over vast landscapes, they tried spraying it on the interior walls and ceilings of homes, field-testing it in the homes of poor Black cotton farmers in Arkansas and tenant farmers in Georgia.

It seemed to work. Mosquitoes died, and, more importantly, malaria rates declined. In 1944 the agency appealed to Congress for funding to spray homes across the nation's malaria belt, a thirteen-state expanse stretching from California to the Carolinas. Congress said yes.

In its first year the agency's Expanded Program, as it was called, sent spray teams into more than half a million homes across the southern United States. Gaining entry took a little convincing. Insecticides were something people were accustomed to using outside, not inside. They were known to be poisonous. And no one wanted sprays marking their walls, counters, tables, pots, pans, or dishes. Most home owners acquiesced—or believed they had to. A photo from the program in South Carolina showed a Black family of fifteen sitting dutifully on their front porch with furniture, clothing, bedspreads, and curtains piled high around them as the crew approached.

In Hidalgo, Texas, the spray teams "learned a lot about the psychology of housewives," said county health officer Mary Walton. They refused to move their furniture, questioned the spraying, and asked the crews to treat barns and outbuildings instead of their homes. But in the end, Walton wasn't too worried about the housewives turning away her crews. Because even though there wasn't much malaria left in her county, the spray killed flies, which, as everyone knew, spread dysentery, typhoid, and polio. Everyone feared polio. And the state was experiencing a summer "siege" of the disease. Health officers were urging everyone to eliminate flies and their breeding places to prevent polio, and the local chapter of the NFIP was in high gear, organizing fund-raisers and collection drives to help polio patients far and near.

•

Polio cases weren't just up in Texas; they were on a "sharp" rise across the US, the foundation's medical director, Don Gudakunst, announced that spring. To focus public attention on the disease in the midst of war, the foundation's publicity department launched a campaign capitalizing on the "Miracle of Hickory." Millions of posters hung in shops and libraries, and brochures mailed to suburban homes depicted the Hickory epidemic's youngest (and whitest) polio survivors: three-year-old Judy, who had regained the ability to walk; paralyzed one-year-old Jerry, who made a fast recovery; and infant Kenneth, fully recovered after polio "crippled" his back and legs. "These are some of the Children your Dimes and Dollars Helped," they read.

The foundation raised $19 million that year, much of it coming in ten cents at a time, a literal march of dimes, from individuals and families moved to donate what they could, not just by the campaign but by the loss of President Roosevelt, who had died suddenly from a brain hemorrhage that spring.

Down in Raleigh, Crabtree was caught up in the foundation's grand ambitions to bring as much attention to polio as possible. He pulled together a report on polio's likelihood of striking North Carolina again and sent it to the local papers. Most people, Crabtree pointed out, thought the state would be spared in the summer of 1945 because it had been hit so hard the summer before—which made sense because so many people had certainly become immune in 1944. But the disease could still return. California, for one, had had three bad polio epidemics in a row. "This is no time to relax," Crabtree wrote. Polio "strikes where it wishes . . . one day a household is well and happy; the next, there is dread and terror."

When Crabtree saw the DDT suggestion in the *Hickory Daily Record*, inspiration struck. Here was something big his state could do to keep polio at bay. He tore out the letter and mailed it to Gudakunst at the foundation in New York.

"Most interesting," Gudakunst replied. He decided to let Crabtree in on a piece of news. The foundation's research department was already planning several "rather extensive" field trials of DDT for that summer. The plans hinged on the federal government having enough pesticide to spare from war and other official use. With the war's end in Europe, he expected this might happen soon. But to carry out the plans, the foundation had agreed to "a rigid secrecy agreement with the Army." Gudakunst asked Crabtree to make some local calls, to drum up a little local support and excitement, all while keeping folks mum for the moment. "When you are talking with the editor and the good Dr. Melzer," he told Crabtree, "you might tell them to hold their fire and not get too much publicity started."

•

That June, the army released some of its DDT supplies to the foundation, and the foundation sent another grant to Yale. Horstmann's colleague at Yale, Joseph Melnick, a thirty-year-old with a PhD in physiological chemistry and an interest in epidemiology, took the lead this time.

To test DDT, they needed to conduct "a *controlled* study," Melnick stressed. He worked with his colleagues in the polio unit to plan all of the details of their experiment. To start, they'd need to divide a city into areas to be sprayed and those to be left untouched. That way, they could compare the fate of flies between the two. He got out a map of Savannah, Georgia, and started to mark it up.

Chapter 3

FLIES

A Bean Company orchard sprayer could look a little like a body missing its head. But not the bigger ones, which had a full-sized cab and a tank that could hold a thousand pounds of spray. Bean sprayers typically rolled through orchards; farmers had been using them since the late nineteenth century to spray lead arsenates on fruits and nuts. But in the summer of 1945, a Bean sprayer rolled down the city streets of Savannah, Georgia, loaded with a thousand pounds of DDT.

One man drove the truck while four men stood on its running boards, wielding hoses that released a steady stream of the pesticide in solution. The emulsion settled on the trees and houses lining the streets. It was all part of Joseph Melnick's plan. Nine days later, researchers from Yale's polio unit trapped and counted flies in the sprayed and unsprayed areas. The DDT had made no difference.

Back in New Haven, Melnick and his colleagues regrouped to figure out what to try next. They decided to adjust their methods. They set stations of caged flies throughout the city and then rigged a Bean sprayer with an "air-blast apparatus" powered by a 100-horsepower motor, which shot 125,000 cubic feet of DDT-containing mist per minute over the city's streets. The caged flies must have died, but Melnick's published report didn't include the results—only the modified methods.

Then the team moved on, making plans to test the high-powered sprayer in a city in which a polio epidemic "appeared to be developing." When a handful of polio cases appeared in the eastern part of Passaic County, New Jersey, that July, the team members divided the city of Paterson into sprayed and unsprayed areas, as they had in Savannah. They rolled the air-blast sprayer down its streets while a team of men with DDT canisters on their backs sprayed people's garbage cans. They documented a "definite fly reduction lasting a few days in all wards." The number of polio cases in the sprayed area was also lower than that in the unsprayed areas: seventeen versus twenty-two. The difference was real, but it wasn't big enough to prove anything.

And then opportunity struck. While the Paterson test was winding down, polio cases in Rockford, Illinois, began making the news. By early August 1945, five dozen Rockford residents had contracted polio, and five of them, all children, had died. The city's health commissioner, N. O. Gunderson, knew the county was short on medical professionals, especially those equipped to treat polio patients. He issued quarantines on children, postponed the start of school and the fall football season, and appealed to the state's board of health and a local chapter of the NFIP for help. Publicly, the foundation pledged its unlimited financial assistance. Privately, its research director called John Rodman Paul's team at Yale to make plans for another test of DDT.

Just as they had in Paterson and Savannah, the Yale scientists divided Rockford into areas to be sprayed and those to be left alone. As they had before, they hoped to keep the aim of the spray a secret. But it took weeks of meetings to secure the support of local officials, and at a meeting where the city's health committee was scheduled to vote on the proposed study, word of it got out. The next morning, newspapers across the state announced that Rockford would be fighting its epidemic with DDT. Within a few days, wire services were spreading the story across the country. The foundation's public relations head, Fred Maguire, dashed off a memo to Gudakunst: "It seems that the story of the Foundation and DDT has leaked." Gudakunst urged him not to let word spread any further.

As reporters kept calling, Maguire feigned ignorance. "It was all news to me," he said.

Worried that the news was going to destroy his team's hopes of a controlled experiment, Paul set out from Yale for Rockford. Polio epidemics, he knew, made people "almost hysterical." It was possible, he feared, that people would "stampede into sprayed areas" from the control areas for protection. He only hoped he could personally convince local politicians, reporters, and the public to let the experiment proceed as planned so that his scientists could prove—or disprove—DDT's actual usefulness against polio in the long run.

But DDT's use against one of the most feared diseases of the day, on US soil, was too big a story for reporters to ignore. When Maguire wouldn't answer her questions, Jane Stafford, a chemist who worked as a journalist for the Science Service, kept calling around. Eventually, she reached the office of the surgeon general, where a captain with the Medical Administrative Corps was all too happy to give her a scoop.

Rockford wasn't just going to be sprayed with DDT from the ground, he told her. The city was going to be sprayed from above by a repurposed B-25 Billy Mitchell army bomber on its way to Illinois from the Canal Zone in Panama. Its bomb bay was already loaded with a 550-gallon spray tank for mosquito control. The story was even better than other reporters knew. Stafford's article held nothing back. "Airplane to Spray DDT on Entire City," it announced.

When Paul arrived in Rockford, he found the city "in an uproar," he said. Local leaders and the public were furious—with him. The notion of an experiment in the midst of an epidemic was unfathomable. A local senator accused the scientists of withholding a cure from people desperate for relief. The federal Public Health Service got involved, urging Paul and his scientists to relent. The best course was to spray Rockford as quickly as possible, officials from the service told him, even though news reports had compromised the study's integrity. Paul believed that he had no choice. He agreed.

Two days later, the olive-drab, star-spangled bomber took off, not from the Canal Zone but from nearby Truax Field in Wisconsin.

Two-inch-tall headlines broadcast the news in the papers: "Professor Expected to Board Mercy Plane at Noon Today." "Preventive Spraying for Polio as Important to Rockford as the Atomic Bomb." Americans had dropped the obliterating bomb on Hiroshima two weeks before.

Paul didn't board the plane—he had never planned to—but the plane did fly low over Rockford's treetops and gabled rooflines at 200 miles per hour, releasing a stream of DDT 150 yards wide with each pass. The flight was still meant to be an experiment, but the next day, newspapers across the country ran a wire photo of the plane's pass and declared that DDT had found its peacetime "mercy duty."

The ongoing press coverage rattled Paul, who struggled to keep his composure. "To say the least, it was not the sort of atmosphere in which to carry out a carefully controlled experiment," he said. Privately, he fumed to Gudakunst that publicity had "ruined" his study. The control areas had all been compromised. The airplane spraying hadn't even worked. Although hundreds of papers across the country covered it as a dramatic technological feat, his team's own investigations found that too little DDT had reached the ground. The plane had applied only a fifth of the DDT applied by the ground crews in Paterson. The researchers had been forced to make a quick decision to repeat the spraying by hand, an effort that took eleven men working day and night six days to complete. By then, said Paul, "We had lost two precious weeks."

Cases were falling, but as the team tallied and graphed the numbers, they realized that the epidemic was in fact already on the wane when the spraying had begun. Deflated, Melnick wrote up their conclusions. DDT had temporarily reduced the flies in Rockford, as it had in Paterson. However, the spraying had "no effect on the poliomyelitis epidemic in either area," he wrote.

At that point, however, the unit's conclusions were nearly beside the point. Once Stafford's story ran, Maguire had decided that the more publicity, the better. "It seems to me pretty logical that we should report to the general public that sizable chunks of their dimes are being expended for this purpose," he said.

The foundation's supporters in the army were all too eager to help spread the message, too. "This is only one of a hundred of ways," an army colonel told the press, "that materials and methods developed by the equipment laboratory will be used for the benefit of the American public in peacetime."

As it happened, news of the Rockford mercy mission hit the stands just days before the War Production Board released DDT for sale.

As DDT rolled onto the market in late August and early September of 1945, cities and towns immediately deployed it against polio. When a high school student in the Texas Panhandle died of the disease, the superintendent sprayed the school and all of its buses with DDT. After a seven-year-old succumbed to polio in Dade County, the county's health director asked the navy to spray Miami, Miami Beach, and Coral Gables by plane. Chicago sprayed the homes of polio-afflicted families, city dumps, and fly-ridden intersections. Boston sprayed garbage cans, refuse piles, and dumps and announced that schools would be next. Richmond, Virginia, announced plans to spray the whole city. In late September, Boston's mayor announced that since certain neighborhoods had been sprayed, not a single case of polio had occurred.

"We are making no claims," he said, "but something significant happened here."

•

Polio receded as always that fall and winter, as the nation slowly recovered from the war. And then the 1946 polio season took off, a month earlier than usual. In May the nation had more cases than it had seen the May before. South Florida was hit hard, with 126 cases and 6 deaths early that month. Horstmann packed her bags and boarded a plane for Miami, again dispatched by the National Foundation; polio was, until then, considered rare in tropical climes, making the city's epidemic a scientific curiosity.

Texas was also counting an increased number of cases again, especially in San Antonio, which by mid-May had nineteen cases and five deaths. The city shut its swimming pools, banned children from churches and streets, prohibited picnics, and ordered all high schools to hold graduation ceremonies over the radio. Then state health officer George W. Cox announced a plan to "eliminate" the housefly and the places where it bred. "Texas has been dirty for years," he said. "It is full of breeding places of this sort of malady and the time has come to clean up the state, thoroughly and at once."

Texas towns and cities without any cases heeded his call. Austin sprayed its schools with DDT, Rockport sprayed all incoming buses and trains, and the Lubbock Health Department rigged a jeep into a spray unit for treating garbage cans. Exterminators, drugstores, gas stations, and chemical companies pushed their DDT. "Fight Polio with DDT," a service station urged people in Pampa. "Kill Polio Carrier Flies," DuPont ads commanded in Corpus Christi. In San Antonio the Sherwin-Williams store invited customers to come down with empty canisters to fill with "Pestroy" DDT for free. DDT sales in the city shot up 600 percent.

But most of the DDT in San Antonio was sprayed by the city, not home owners. The city health department added sixty men to its team of fifteen "sanitarians," handing them equipment to spray residential garbage and San Antonio's twenty thousand outdoor toilets with DDT. The mayor's office contracted with a Houston-based DDT manufacturer, which sent nine army planes to coat the city's garbage dumps and "swamps." City health commissioner H. L. Crittenden then got an offer from New York–based Todd Shipyards Corporation that he couldn't refuse. The company had built fog machines to generate smokescreens for ground campaigns in the war, and they wanted to ship one to San Antonio. In tests on New York's Jones Beach, Todd engineers had shown that the foggers could generate a fine DDT mist. The company hired photographers to capture children playing in it and a bikini-clad woman clouded in it as she had a hamburger and a Coke. The mist was safe, Todd promised, and so delicate that bees would survive in it even as flies perished.

Todd was so eager for San Antonio to put the fogger to the test against polio that the company offered it to the city for free.

Crittenden couldn't say no. Within a week, the smokescreen machine was rolling down his city's streets, its DDT fog rising gently and coating everything in sight. The health department told city residents and business owners to leave their windows open so the mist could settle outside and in.

San Antonio's epidemic peaked in June, waned in July, and was over by early August. Close to a hundred people were infected; a dozen died. Crittenden's department drafted a report on the decline in cases that followed its DDT-based anti-fly campaign and sent it to a scientific journal. San Antonio was the only city, it claimed, to use "all known forms" of DDT against polio. The department believed it had proved that a polio epidemic could be slowed in the height of summer with widespread DDT. "The DDT treatment," one health officer said, "was an answer to our prayers."

•

Back at Yale, Paul was still managing his polio unit, fuming over the mess in Rockford and frustrated with the ongoing use of DDT. Unwilling to study the insecticide any further, he asked Gudakunst if he could spend leftover grant funds on an electron microscope instead. "A microscope is not packed with as much dynamite as is the spraying of DDT," he bargained.

Paul's Yale colleague, Charles-Edward Amory Winslow, then editor of the *American Journal of Public Health*, was equally aghast at persistent attempts to fight polio with DDT. The approach had "little scientific warrant," he said. It reminded him of public health "a century ago, when people were driven by blind fear, ignorance, and superstition." His words stung health officials in San Antonio—and they struck back. The "Yale man," said a city health officer, had made "one of the most out of date statements I ever heard."

In the polio seasons that followed, it was the view from San Antonio that prevailed. In 1945 a couple of polio cases were grounds for shutting down public life. But after the summer of 1946, DDT kept America open for business even as polio spread. State fairs, parks, beaches, and playgrounds welcomed crowds, local officials proudly announcing their treatment with DDT. Counties didn't wait for a case to appear before spraying schools, orphanages, and penitentiaries. In small rural towns, crop dusters sprayed main streets and their surrounding neighborhoods to keep polio at bay. In tonier areas, foggers rolled across golf courses, racetracks, country clubs, and resorts to prevent disease. From New York to California, children played in the mist. In the polio years a cloud of DDT seemed one of the safest places to be.

But municipal enthusiasm for DDT put the National Foundation in a bind. Its grant-funded scientists, such as Paul, dismissed it. University of Michigan virologist Tom Rivers, whom the foundation had contracted to guide its research agenda, denounced it. Spraying DDT to fight polio, said Rivers, did "about as much good as spraying soapy water from planes for personal cleanliness." Yet the foundation's donors—the countless men and women who volunteered their time and money on behalf of the thousands of local March of Dimes chapters across the country—were all desperate to use DDT. To an extent, Paul sympathized. The call of the polio years was to "try something—do anything!" he admitted. And DDT seemed to answer it.

The foundation knew that its own publicity juggernaut had lifted public hopes for DDT. For several years after the Rockford epidemic, foundation research director Harry Weaver kept careful track of where DDT was sprayed and where polio appeared. The pattern proved Paul right. The health director of Wilmington, Delaware, sprayed the city with DDT every two weeks for two years and then suffered an outbreak in 1947. An epidemic struck the virtually flyless town of Caldwell, Idaho, that same year. Hidalgo County, Texas, spent three years using DDT to suppress flies and then suffered an outbreak in 1948.

DDT, the foundation learned, simply did not combat polio. After the Hidalgo epidemic, Weaver attempted to draft a statement saying as much so he could distribute it to journalists and local chapters asking about DDT. But he didn't quite know what to say without embarrassing the foundation itself—not to mention the DDT-spraying officials and communities all across the US that the foundation depended on for support. So he chose his words carefully. All epidemiologists agreed that flies could potentially transmit polio, he noted. But most believed DDT could not *prevent* polio. And even if flies could transmit polio, he added, they didn't transmit most polio. This meant that an epidemic couldn't be stopped by spraying DDT. He mimeographed the statement and sent it off.

Down in Orlando, some of Edward Knipling's former colleagues wondered whether this widespread DDT use might have a downside. Entomologists knew well that insects could evolve to withstand a chemical poison their species was repeatedly exposed to. Knipling himself had shown that screwworms exposed to the insecticide phenothiazine over several generations became resistant to it.

So two of the scientists still working out of the Orlando field station devised an experiment. They put 300 houseflies into a 100-cubic-foot box and then filled it with DDT/kerosene spray. A small number of flies survived. They used those flies to breed a subsequent generation, which they then placed in the box and sprayed again. They repeated the process fourteen times, each time producing a new generation of what they called "special" flies, more likely to survive a hit of DDT than "regular" flies.

Journalists who wrote about the study dubbed the survivors "Super Fly." By the time Super Fly made the news in 1948, the Orlando lab was getting reports of resistant flies from all across the country. Communities that had been able to kill all of their flies with DDT in 1946 were seeing more and more of them survive.

The Orlando scientists, now up to the fifty-fifth generation of Super Fly, went into the field and collected flies from dairies and garbage dumps in Apopka, Winter Garden, and Kissimmee, all towns near Orlando. DDT, once lethal to just about every fly, killed only about 11 percent of them now. Similar reports were coming out of Europe, too. Up in Atlanta, the MCWA, renamed the Communicable Disease Center (CDC) to reflect its growing postwar responsibilities, reported that in its own tests more than 75 percent of houseflies were surviving DDT.

•

That same year, CDC scientists were studying the influence of fly control over diarrheal disease in Hidalgo County, Texas, when five cases of polio appeared. The area, a flat, dry portion of the lower Rio Grande, had never experienced anything other than sporadic cases of the disease. At first, the cases raised no alarms. But by May, there were more than 150 cases across a three-county area. The area was already being divided into sprayed and unsprayed areas for the diarrheal disease study, so the scientists decided to keep track of the polio cases in each.

When the epidemic subsided, the scientists found no difference in polio cases or the timing of their onset in the treated and un-treated areas. But all of the people who came down with polio, they found, had come in contact with people who had paralytic cases of the disease. Polio, the scientists concluded, wasn't spread by flies but by people, especially those with severe disease. The study findings, commented Paul, were "a final judgment" on the whole question of fighting polio with DDT.

The year after the Hidalgo outbreak, in January 1949, the World Health Organization made an official announcement: the "common house fly" was officially "immune" to DDT. But it took the NFIP another year to take a firm stand on DDT's use for po-lio. In March 1950, Weaver drafted an updated statement for the foundation's local chapters: "THE NATIONAL FOUNDA-TION DOES NOT SANCTION THE USE OF CHAPTER FUNDS FOR THE PURCHASE OF DDT SOLUTION OR THE HIRE OF PERSONNEL AND/OR EQUIPMENT FOR THE DESTRUCTION OF FLIES IN CONNECTION WITH ATTEMPTS TO CURE POLIOMYELITIS OUTBREAKS." The foundation was officially done with the chemical.

And Weaver was right to assume that lowercase letters wouldn't get the message across. The Hidalgo study may have been the last word on DDT and polio for scientists and the foundation, but the rest of the country let go slowly. From Anniston, Alabama, to On-tario, Oregon, at the first signs of an epidemic, towns continued to spray. When polio hit San Angelo, Texas, in 1949, city officials acknowledged that DDT would do little to stop the disease and would likely make local flies resistant. On the other hand, they reasoned, spraying made for "a more pleasant place" and "helped calm persons who had become panicky over polio."

So they assembled a DDT spray crew and sent a fog machine down the city's streets.

Chapter 4

PRODUCTION

All the workers at the Rothbergs' chemical plant knew that the family ran a tight ship. The plant sat on the first bend of the Passaic River, just north of the Port of Newark, where manufacturing facilities lined up along a thick cord of railroad lines. Its workers churned out large batches of an industrial chemical called TCP, the herbicides 2,4-D and 2,4,5-T (chemicals later combined to make Agent Orange), and—more than anything else—DDT. Ben Rothberg walked the concrete floors of the single-story building several times a day. He looked for spills and leaks. He made sure no one was sitting idle. And he peered inside every trash bin, making sure it contained no residue or compound that could be salvaged and fed back into the production chain. Neither he nor his father, Pincus, tolerated such waste.

Long ago, before the war, his father's specialty had been drugs. After graduating from the College of the City of New York in 1912, Pincus Rothberg worked for the federal Bureau of Chemistry in New York, analyzing the "patent" medicines sold at city drugstores to make sure that they contained only their listed ingredients and nothing poisonous or illegal. His big moment came when he discovered strychnine in Dr. Williams' Pink Pills for Pale People. He joined the American Druggists' Association, moved up its ranks, and then figured out a way to make aspirin that didn't upset the stomach. He patented the process, leased a plant in Newark, and started production. He had a wife by then, Rachel, and two young children, Miriam and Ben. He called his firm the Montrose Chemical Company.

Montrose made aspirin until World War II, when the War Department (he'd done some work for it during World War I) asked him to shift some of his production over to chemicals for the army, including TCP. Without hesitation, he did. When, two years later, the War Department needed DDT, Rothberg shifted course again. With fewer than ten workers, Montrose was smaller than most of the other wartime DDT producers, such as Monsanto and DuPont. But for Rothberg, that wasn't a problem; making DDT was no more complicated than making aspirin.

At his little Newark plant, his company began to make something called technical DDT, a mixture of several DDT isomers, molecules that differed in the location of their chlorine atoms. Technical DDT was mostly composed of an isomer called p,p'-DDT, or "para, para prime" DDT to chemists like him. A molecule of DDT contained five atoms of chlorine, three of them attached to a single carbon atom, the other two each holding on to their own carbon ring. In p,p'-DDT, the independent chlorine atoms were equidistant from the cluster of three attached to the carbon atom. This was the isomer that spelled death to so many bugs.

It took two key ingredients to make technical DDT: monochlorobenzene, a carbon ring with a chlorine atom attached, and chloral, a fairly simple molecule with a lot of chlorine atoms. (On its own, chloral had a dark side: combined with water it became chloral hydrate, a powerful sedative known as a Mickey Finn.) The Montrose plant made its own chloral by adding chlorine to ethyl alcohol and distilling the resulting compound to remove water because water slowed production. Once the chemists had pure, dry chloral, they dissolved it in monochlorobenzene and then poured the resulting solution into 2,000-gallon reactors at the back of the plant. In these vast vessels, the reaction between the chloral and the monochlorobenzene produced water, heat, and DDT. Plant workers added another chemical, oleum, to absorb the water. They kept the reactors refrigerated, encased in cooling jackets and ice, to control the rate of the DDT-producing reaction. When the reactors contained nothing but DDT, the workers rinsed it with hydrogen peroxide to wash off any impurities. They then poured the washed DDT into large, shallow pans, six feet wide and ten feet long. Each pan held two thousand pounds. They let the DDT cool. As it cooled, it crystallized. If the government wanted flakes, the workers chopped the crystals into lumps that looked like sugar candy. If the government wanted powder, they ground it down to dust. They then packaged it up and shipped it out. It all smelled faintly of fruit.

The Rothbergs didn't like waste, but some waste in the course of production was inevitable. The reactions in the refrigerated vessels produced acidic by-products in addition to DDT. Plant workers drew off some of the acid, carting it back to the front of the plant to dehydrate more chloral. The rest they let drain into a sewer that carried it to the river. The hydrogen peroxide rinse also needed somewhere to go, so it went into the sewer as well, each time taking with it a little bit of DDT.

During the war, Pincus Rothberg had sat on the War Production Board's DDT advisory committee, the one that had voted to sell the pesticide to the public. And when the war was over, his company kept right on making technical DDT. People seemed to need it as much as they needed aspirin—maybe more. Now, a year after the war, his customers included the federal government, local governments, and pesticide formulators that included DDT in everything from insect bombs to vacuum attachments. His business was booming. He was hiring more workers. He had named Ben, now twenty-six, vice president. And when Miriam married an accountant named Samuel Rotrosen, Rothberg put him in charge of the company's finances.

Now Rothberg was eyeing the market west of the Rockies, where he knew that no one was producing commercial-scale DDT. He decided to reach out to one of the more established West Coast chemical firms, the Stauffer Chemical Company. It was a family business like his, but much bigger, based in San Francisco and founded in the late nineteenth century. It operated plants up and down the California coast, with a handful of new plants going up in Nevada. Rothberg approached its executives with a proposal. For $100,000 he would give them 25 percent ownership of Montrose and invest the money in construction of a West Coast Montrose DDT plant that would procure its raw materials from Stauffer. Stauffer countered, offering $50,000 for 50 percent ownership of Montrose, but also offering to build a new plant in the industrial town of Henderson, Nevada, which Montrose could lease to produce chloral and monochlorobenzene. With Stauffer's connections, Montrose could then truck the ingredients to Torrance, California, where the two companies could build a jointly owned DDT plant on property owned by Stauffer and leased by Montrose. Out west, with two plants to run, Montrose could make far more DDT than it had ever made in Newark.

Rothberg said yes. Agriculture in California was flourishing; he knew that with this plan he could corner the western market for DDT. He sent one of his Newark engineers, a chemist named Max Sobelman, to Los Angeles to get the DDT plant up and running. The Stauffer site was roughly fifteen miles south of downtown LA and six miles east of the massive and bustling pier at Redondo Beach. The plant, he knew, would need the same reactors and sheet pans as the one in Newark, but also a collection of drainage ditches to catch the acid sludge and hydrogen peroxide rinse. Sobelman got to work.

•

Down in McIntosh, Alabama, Geigy's new DDT plant went up in the middle of an industrial boom for the town. Oil prospecting had turned up a giant salt dome, and the Mathieson Chemical Company had set up a plant that turned the salt into chlorine and caustic soda. Insecticide makers rushed to set up shop nearby to turn the chlorine into organochlorine insect killers. Geigy made DDT. The John Powell Chemical Company manufactured DDT, toxaphene, and hexachloride. The California-based Calabama Chemical Corporation set up a plant to manufacture monochlorobenzene and then trucked it to a plant the company leased from the US Army on the army's Redstone Arsenal up in Huntsville. The army had manufactured poison gases at the plant during the war; Calabama now leased it to produce DDT.

The Huntsville plant was a low-slung building topped with belching stacks that sat at the center of a sprawled complex of sheds, steel towers, and holding tanks. A string of power lines ran along one side. Down at ground level, a pair of brick-lined ditches lay ready to capture whatever emerged from the plant. Calabama started production in 1947. The plant ran seven days per week, churning out up to 2.5 million pounds of technical DDT per month. Every day, it discharged more than two Olympic-sized swimming pools' worth of wastewater into the ditches.

When the ditches filled, they overflowed into a long, wide gash in the earth, cut in a series of cascading levels, that carried the cloudy wastewater half a mile down to the nearest open water: the Huntsville Spring Branch. The creek flowed across the arsenal and down into Indian Creek, a larger tributary that met up with the wide Tennessee River at the tiny town of Triana. The creeks and the Tennessee in that part of Alabama were all part of the Wheeler National Wildlife Refuge, a sanctuary created by President Roosevelt back in 1938. Tens of thousands of migrating ducks and geese overwintered in its sloughs each year, flying in from as far as northern Canada. Great herons and little blue herons fished in its waters. Bald eagles perched in its trees. American alligators rested on the banks of its tributaries, which teemed with catfish, smallmouth buffalo, largemouth bass, carp, shad, and bream.

In 1948, about a year into production, a Calabama plant manager noticed that DDT, insoluble in water, was settling at the bottom of the brick-lined trenches outside the plant. A utilities chief at the arsenal later noticed the same. Neither seemed too concerned.

Then Calabama's McIntosh neighbor, Mathieson, merged with the arms and chemical manufacturer Olin Corporation to create the Olin-Mathieson Chemical Company. Olin-Mathieson moved quickly to acquire Calabama. And with that, Olin-Mathieson controlled the entire production chain, from the chlorine in McIntosh to the technical DDT in Huntsville. When the crystals formed in

Huntsville, truckers drove them back to McIntosh, where they were crumbled or pulverized and then bagged and sold to insecticide makers, some of them just down the street.

Sometime in 1955, some of Olin's managers and engineers noticed a white substance making its way to the Huntsville Spring Branch. They decided it wasn't DDT. Not long after, army officials at the arsenal asked Olin to make sure that no DDT was making its way into the creek. They told the company that although the creek cut across the arsenal, it was still part of the federally protected wildlife refuge. Decades later, Olin officials would deny that the army ever said a thing.

•

Montrose's Torrance plant was also up and running by the spring of 1947. But in truth, the timing wasn't ideal. DDT manufacturers had produced a surplus in 1946, selling just 36 million of the roughly 50 million pounds they produced. As a result, the pesticide's price was dropping.

One by one, the big manufacturers got out of the market. DuPont, for one, had never made any money on DDT; off patent, it was among the cheapest insecticides out there, and the profusion of manufacturers kept pushing its price ever lower. Eventually DuPont was actually losing money: three cents for every pound it produced by the early 1950s. So it dropped DDT. So did Hercules, Monsanto, and Merck. If they could develop their own proprietary insecticides, they all figured, they could make a lot more money.

But Montrose stayed in. DDT's falling price was a problem for Rothberg, too, but he wasn't prepared to leave the market altogether. Instead, he took a gamble. In 1953 the company suspended all DDT production in Newark, transferring it all to the California plant. Demand held, and then—as the other manufacturers dropped out—it grew. Production at the Torrance plant steadily rose.

But the waste ditches couldn't keep up with the volume. In the early fifties, Montrose switched from batch to continuous production, and though they tried to send their liquid waste through the county storm drains, it routinely overflowed. The acid sludge, meanwhile, was too thick for the drains. Composed of sulfuric acid, organic substances, and a tiny amount—no more than 1 percent—of DDT, the viscous brew was coming out of the plant at a rate of more than 10,000 gallons a day. Workers eventually began collecting it in holding tanks and barrels. A local disposal firm, the California Salvage Company, loaded the sludge onto barges at the Port of Los Angeles, carried it out into the Pacific, and dumped it in the sea. The rest of the plant's waste went into nearby landfills.

The Montrose factory site itself was messy, piled high with waste, and local residents began to complain to county officials. The Los Angeles Department of Public Works regularly inspected and asked the company to apply for disposal permits. The company complied. It had a permit to send its liquid waste through the sewers to a disposal plant at White's Point, roughly twelve miles south, just outside of coastal San Pedro on the Palos Verdes peninsula, which discharged into the Pacific. And it sent its sludge to permitted landfills and ocean dumps.

In the meantime, the plant's volume continued to grow. By 1956, Montrose was making 40 percent of all DDT manufactured in the United States. The Torrance plant was worth $16 million. Stauffer offered to buy Montrose outright. The Rothbergs and Pincus's son-in-law, Rotrosen, declined. The decision put the company on course to become one of the country's last defenders of DDT, a role that Rotrosen would play decades down the line.

Chapter 5

ECONOMIC POISONS

Claxton, Georgia, was a classic small town in the late 1940s, with two soda-fountain drugstores, a grocery, a hardware store, and a bakery famous for its pecan-raisin fruitcake. The Nancy Hanks II streamliner train passed through twice a day, carrying its segregated passengers between Atlanta and Savannah. Highway 280, newly paved, cut through town parallel to the tracks, connecting Savannah to the east and Americus to the west, lined with small farms in both directions.

Doffie Colson was new to Claxton, one of millions of Americans whose lives were uprooted by the war. Her own Methodist family's roots ran deep in rural Tattnall County, just to the west. But when Congress authorized the army to build dozens of new camps and training centers for the war effort, her family and hundreds of others—more than half of them Black, the rest white like the Colsons—were moved off their land to make way for a new antiaircraft artillery range. It was a "forced exile," said Colson's sister. But they gave up their homes and, "most valuable," their neighbors, Colson said, "to help train soldiers in the nation's hour of need."

Their new Claxton farms were small but big enough to be self-sustaining. Colson farmed and raised bees with her two daughters. Her sister's family, on the farm next door, grew vegetables and truck crops and blooms that won the local flower shows. Highway

280 gave them access to Georgia Power, phone lines, and passenger buses right outside their doors.

All told, Colson thought they had found a promising place to live, until the spring of 1945—three years after they moved in. Colson fell sick with a sore throat, "nervous chills," and muscle pain so intense she sometimes took to bed for weeks at a time. Her sister, Mamie Plyler, fell sick too, her mouth and throat erupting in painful sores and her head aching with a persistent "irritation" that didn't give way for months.

A local doctor diagnosed Colson with allergies, but his treatments didn't work. Colson found a specialist up in Atlanta, but his treatments failed too. Then her daughters began to feel sick, and Colson was at a total loss. But her sister had a theory: Plyler wondered if their health problems might be related to the planes that, since the war, had been using Highway 280 as a runway to dust the vast tobacco, tomato, and peanut fields owned by big landowners near their farms.

Sitting in the kitchen of her little farm home on the highway, Colson thought about her sister's theory as she read the farm papers, where she learned all about the new poisons developed since the war. The papers had much to say about DDT, which killed mosquitoes, flies, boll weevils, bud beetles, corn borers, hornworms, tobacco worms, and more. Compared to the older poisons, such as lead and arsenic, DDT was a "wonder drug," said the farm paper *Capper's Weekly*. But it was also "fussy stuff" that farmers had to use with care. Channing Cope, Georgia's popular farm columnist, was impressed with DDT but also nervous: because it killed pollinators, he said, he feared it had "the power to ruin us."

Colson, like any farmer, knew that the key to using poisons had always been to use just the right amount, right where it was needed. When poisons rained down from a crop duster, however, that was just about impossible. So when the crop dusters flew low over their home, she closed their windows, brought in the animals, and took down any laundry on the line—if she could get to it all in time.

Then one morning in the spring of 1948, a crop duster parked near the entrance to Colson's property to set up for its flight. As the pilot loaded the hopper, a wind picked up and blew the plane's load toward the driveway, where her older daughter, Dorothy, stood waiting for the bus to school. Without warning, the girl was bathed in a fine dust. She ran inside. Before her mother could clean her, she began vomiting and grew weak. Her temperature climbed, and for days she suffered a high fever and severe sore throat.

Colson now knew her sister was right. The illness in her family wasn't allergies. It was "straight out poisoning," she said.

•

Colson was a practical woman, petite with a bun and a heart-shaped face. After her daughter recovered, she decided not to plant her usual vegetable garden by the highway anymore. Poisons, she now felt sure, would make anything she grew "unfit" for use. She devoted the time to writing instead, sending letters to anyone she could think of to ask for help figuring out how to stop the planes from spreading their poisons on her family and their land. Colson had never finished school, but in her careful, sloping script she wrote to the Georgia Highway Patrol, the Civil Aeronautics Association, the Georgia Department of Commerce, and the Georgia Department of Public Health, to start.

At the health department, her letter landed on the desk of Guy Lunsford, director of the department's Division of Local Health Organizations, who sent her a reply. There was nothing he could do to help, he told her, because any action the state might take would have to be statewide, and it just wouldn't be possible to halt dusting at that scale. He suggested that she try contacting her local health commissioner and that she send her soil to the Georgia Division of Industrial Hygiene to learn exactly what sprays it contained.

Colson collected some soil and sent it off. And with Lunsford's help, she convinced a man named W. D. Lundquist, the health commissioner for her county, to come through Claxton and see the issue firsthand.

Lundquist was a health commissioner with an avid golf hobby, who made it into the local papers more often for his game than his job. On a baking-hot Friday that July, he drove through Claxton, stopping in first to see folks he knew. A local doctor told him Colson's daughter had probably had strep, not poisoning. The landowners who held the acreage near Colson's farm told him they sprayed something called Vapotone, especially on the tobacco, but that it deteriorated too fast to harm anyone caught in a plane's drift. He stopped in at the county agent's office, where he learned that DDT was sprayed over the nearby town of Daisy; it was one of dozens sprayed by the state to combat an infestation of white-fringed beetles in shade trees. Not one of the hundreds of thousands of people in the sprayed areas had complained, the agent said. Lundquist wasn't surprised. His own office was dusting rat runs throughout Evans with DDT for typhus control, and no one had complained to him, either.

When Lundquist arrived at Colson's farm, she showed him where the planes sometimes loaded their poisons by her driveway gate. She showed him letters she had received from the Department of Commerce and the State Patrol, stating that the planes might be violating local laws. She shared cautious-toned articles about the new poisons she had clipped from farm papers, most of them about DDT. She also showed him letters she had received from manufacturers she had written to, all of them urging caution with their sprays.

Lundquist was taken aback by all she had compiled. But he was already convinced that her problem had nothing to do with crop dusting. All of the poisons used near her home were "very mild," he told her. Her and her family's symptoms, he said, must have some other cause.

Colson insisted the poisons weren't mild. She told Lundquist that the symptoms weren't limited to her family and that they also weren't limited to people. Her milk cow was sick, her chicks were dying, and her honeybees, such an important part of her livelihood, were dying too. That, she knew, was definitely caused by DDT; all the local beekeeper associations she had checked in with said so. And "any poison strong enough to kill or damage honey bees," she told him, "is surely strong enough to affect people."

Lundquist wasn't swayed. Insect-killing dusts and sprays "might be irritating to some," he granted, but they could not cause the extent of problems she claimed.

Colson was surprised that a government health official could know so little about the potential harm of poisons. The moment he left, she knew she couldn't count on him to effect any change. So she picked up her pen and wrote again to Lunsford, the state health officer who had helped her persuade Lundquist to visit in the first place. "I appreciate him driving out, as per your request, but must say I was disappointed," she wrote. "I do not think he is a qualified man to deal with such problems as we have here."

Back at his office, Lundquist typed up a report of the exchange from his own perspective. "I tried to the best of my ability for nearly three hours to convince Mrs. Colson that the symptoms she was complaining of could be attributed to other causes," he told Lunsford. "I apparently was unable to change her attitude in the least." The real problem, as he saw it, was quite simple. The symptoms were all "imaginary." In his view, Colson and her sister were merely "overanxious."

•

When Lester Petrie, the head of the Georgia Division of Industrial Hygiene, heard about the poisoning problem in Claxton, it caught his interest. Petrie, an affable northerner with degrees from Harvard and Johns Hopkins, was responsible for investigating the problems posed by chemical encounters in workplaces. Yet ever since the war, his division's work had become more complicated. Before, the job largely involved making sure that people working in factories didn't come in contact with chemicals or equipment in ways that harmed them. Now he was hearing more and more about chemical poisonings in people's homes and farms. Most of them were caused by what those in his line of work called "economic poisons": chemicals designed to kill pests that otherwise caused high-cost damage.

Petrie knew that some 900 new chemical insect killers had been released for sale since the war and that they were now included in more than 45,000 new products that anyone could buy at a feed store or pharmacy. Yet no one knew exactly what kinds of harm they were capable of. For the past few years, Petrie had been compiling a harrowing collection of poisoning cases. Most were crop-dusting pilots choked to death by their own cargo. One was a woman who died after eating blackberries by a sprayed cotton field. Another was a tenant farmer found collapsed in a sprayed tobacco field. A ten-year-old died after taking a taste from a bottle he found on his farm. A six-year-old died after he spilled "plant spray" on his leg. Two children died after making mud pies with "fruit spray." It troubled him that the new poisons seemed as deadly as the old ones and that people clearly didn't know.

Petrie decided it was his job to warn the local public about the specific dangers of the major new poisons on the market. He pored through scientific journals, hoping to learn which chemicals caused which symptoms, or deaths, but he couldn't find what he needed. He wrote to the army's industrial hygiene lab and the USDA, asking if they could share whatever safety information they had. They claimed to have nothing. If he wanted this information, he realized, he was going to have to produce it himself. He asked one of the industrial hygienists on his staff to compile a list of the biggest insecticide manufacturers. One, the Hercules Powder Company, made toxaphene. Two, Hyman and Velsicol, made chlordane. Four others made parathion. Fifteen made DDT. Petrie had his staff reach out to all of them, asking them to share toxicity details about their products. The best he could do in the meantime, he figured, was investigate local poisonings himself and warn farmers to avoid the implicated chemicals.

When Petrie heard about the complaints in Claxton, in fact, he had just finished drafting a warning about the economic poison tetraethyl pyrophosphate (TEPP). His research left him unsurprised that children had died making mud pies. TEPP, sold under several brand names, including Vapotone, was so potent that a splash could kill an adult. A mere drop could cause salivation, lacrimation, vomiting, diarrhea, tightness in the chest, respiratory distress, tremors, and death. It was hard to believe something so poisonous was so easy to get, so he thought it best to have someone check his work.

Petrie sent a draft of his TEPP warning to the FDA's pharmacology chief, Arnold Lehman. In his reply, Lehman praised him for compiling a "timely appraisal of the dangers" of TEPP. He sent a copy to the chief of industrial hygiene at the US Public Health Service in Washington, who said the same thing and encouraged Petrie to tell people to use it very carefully. TEPP was as toxic as he feared, Petrie thought. It was entirely possible that it was the culprit sickening the farmers in Claxton.

So Petrie sent Colson's soil sample to the Georgia Department of Agriculture for analysis because his division had no experience testing for the new poisons. And he wrote to Colson to say that in the meantime, he'd be happy to meet with her the next time she was in Atlanta.

That summer, Colson stopped in. Petrie had no soil results to share yet, so he simply listened to her story. She described mounting medical bills, a steady decline in her farm income, and a growing sense of frustration. "I have tried to be very nice to all who are interested in using these poisons in an effort to save crops," she told him. "Yet when I find they are affecting the health of myself and my family, I feel I am justified in wanting to know more about them."

It was a sentiment Petrie shared, so he promised to make some inquiries. He honed in on TEPP to start. But as he asked around Claxton, he learned that no one was using it anymore. He also became skeptical the more he learned. It was true that a drop of TEPP could kill, but it was also so volatile that it began to break down as soon as it came in contact with the open air. If it had been released from a plane near the Colsons' farm, it would most certainly have decomposed before it could harm anyone below. "It is obvious," he told Colson, that her family's illness "cannot be due to this application." He would look into other insecticides, he told her, but he thought it best if they waited for the soil analysis before he did.

Colson had been elated to have Petrie's sympathetic ear. So it was hard to hear that he had so quickly discarded his theory. When she shared her disappointment with Plyler, her sister pointed out that Colson kept asking outside experts to tell her what was going on in her own backyard. But the two of them were in a much better position to figure that out. We can do this ourselves, Plyler told her. We just need, she said dryly, to use the "tactics of the FBI."

Together, the sisters began to surveil the highway and their farms after the crop dusters left, looking for remnants of packaging blown away or left behind. They struck up conversations with the pilots. They intercepted the insecticide salesmen who pulled into town. The salesmen were a wealth of information, always happy to chat and share company brochures. "These men," said Plyler, "are more concerned over this than our state officials are over protecting the life and health of Georgians." The proof was in the brochures, full of warnings about exposure to their products.

Among the packaging scraps they found, Colson and Plyler identified a bag of parathion and an empty container of chlorinated toxaphene. Another container was so chewed up by cattle they couldn't decipher what it had contained. A pilot told them a neighboring landowner was using something called Vapotone. A salesman told them that the area's tomatoes were dusted with copper and arsenic, and that the peanuts were dusted with DDT. An entomologist for the state told Colson that for the last two years, the state had been spraying nearby Daisy's shade trees with DDT to kill white-fringed beetles. Colson had seen those very planes turning over her farm, right above her chicken yard and, once, over her cow pasture. She believed she had what she needed to follow up with Petrie again.

The source of their troubles had to be DDT, she told him. She had learned even more about it in the farm papers and by writing to other state agencies. It killed bees and chicks, and it stored up in the fat of cows and then passed into their milk, she told him. And now that she knew that both the state and neighboring landowners were using it, she was sure it explained everything she had experienced. She asked if he might urge the state, at least, to spray its DDT with ground crews instead of planes. It would relieve "a lot of suffering," she said.

But Petrie saw little need to act this time. He knew DDT was widely sprayed—he also knew it was beneficial and safe. Colson's county was among those sprayed as part of the federal extended program for malaria control. His own agency, like Lundquist's, had been spraying and dusting it to combat typhus. He composed what he thought was a helpful reply, sending Colson references to articles about DDT in top medical and public health journals. He told her about the people dusted with DDT in Naples during the war, all of them better off for it. "I cannot believe," he wrote, "that the air plane dusting with DDT constitutes any serious threat to human life or health."

Colson received his letter with disbelief. Georgia's state game and fish commissioner had warned that DDT poisoned bees, fish, and birds. Georgia's Extension Service had warned against using DDT on fruit trees because it poisoned pollinators. The state entomologist had warned against spraying it on food crops because it lasted so long and the effects of consumption were not yet known. Yet the health department—alone, it seemed—did not believe that it caused harm. "What a pathetic situation where so little value is placed on human life and suffering," she wrote to Petrie. "And what a dangerous problem it can easily become because those who should have the true facts about DDT (and other poisons) at hand have failed to do this."

Petrie's letter was still fresh in her mind when Colson came across an article in the January 1949 issue of *Collier's* magazine. "The Shame of Our Local Health Departments" argued that health departments provided a critical service but that the nation had far too many of them and that too many were either understaffed or incompetent. "As a taxpayer, it owes you definite services," the article stated. "You should see that you get them."

The article recommended that readers with substandard health departments contact an organization called the National Health Council in New York City. Colson wasted no time. "I am wondering if you could be of assistance in our particular case," she wrote. "We have a health department, that is true, but I am sorry to say one that has refused to give any assistance whatever to what has become a serious health problem." She hoped the council could provide her with some "help in starting a real health movement in a community."

•

Petrie heard little from Colson that winter, while the crop dusters stayed in their hangars. But an article in the *Atlanta Journal* that March caught his eye: "Milk Poisoned by DDT Spray, Farmers Warned." He began to wonder if Colson had been right.

Chapter 6

VIRUS X

By most accounts, the epidemic started with a couple of cows grazing in the southwestern Texas borderland of Big Bend. Open sores and scabs pocked their lips and mouths. A watery discharge slicked from their eyes. Their skin grew thick and rough, their muscular coordination slipped away, and, struggling to walk, the sickest ones extended their front legs and tumbled to the ground, rigid.

The affliction, dubbed X disease, spread quickly, turning up in New Mexico and California next, and as far north as Iowa and Minnesota. In some outbreaks it killed 4 percent of the herd; in others it killed 80 percent. By 1948, a USDA agency veterinarian announced that "no section of the country was free of it."

Miles from any cattle, on the Upper West Side of Manhattan, physician Morton Biskind noticed that a growing number of patients were coming in with a strange cluster of inexplicable symptoms. They all had stomach pain with nausea or diarrhea. Most had a persistent sore throat, joint pain, weakness, and fatigue. A good number were dizzy, giddy, and headachy. In some the joint pain and muscle weakness were so severe they felt paralyzed. Others said the giddiness and headaches were so intolerable they thought they would die.

Biskind wasn't the only one noticing a new illness. Enough cases appeared in California that the state's health department began an investigation. Epidemics of it swept through Austin, Miami, and a college in Pennsylvania. Bing Crosby came down with it after a golf tournament. Errol Flynn's case landed him in the hospital. And it reportedly killed the head of the New York City Police and Fire Department.

No one knew what it was. Some doctors thought it was the return of a gastrointestinal disease first identified in the late 1920s, when it was given the name Spencer's disease. Others argued that it was a new form of the flu or an "intestinal" flu. Still others believed it was an unfortunate combination of "alimentary disturbance" and the common cold. Absent any medical consensus, people began to call it Virus X.

Biskind, a thoughtful man with thick-rimmed glasses that echoed his thick wave of hair, read up on what people were saying about the disease. None of the theories seemed quite right to him. The flu caused fever, but his patients' temperatures were largely normal. The flu, colds, and upset stomachs all resolved fairly quickly, but his patients' symptoms persisted for months. Sometimes their symptoms faded and then returned. Some patients were still sick after a year. He began to suspect that Virus X might be caused by what he called an "intoxication" rather than an "infection." As a gastroenterologist and endocrinologist, he was familiar with both. So he decided to carry out his own investigation.

He began by noting the affected patients in his practice and making sure he had detailed case histories for each one: date of symptom onset, duration of symptoms, source of relief (if any), age, gender, occupation, and so on. He was up to two hundred patients when, as he read through a scientific journal, an article caught his eye. It mentioned that the insect-killer DDT, known for its role in the war, was now widely used in every aspect of agriculture. He knew remarkably little about DDT, other than what he had read in the papers, but the article made him wonder about the possibility of a connection between the poison and his patients. So he dove into the medical literature, looking for reports of DDT's toxicity. They were not difficult to find, although many of them dated a few years back, to the war. He also noticed that more of them were published in

British medical journals than in American ones. But the symptoms they described were eerily familiar.

A lab worker who had soaked his hands in DDT-acetone solution suffered heavy, aching limbs; weakness; and involuntary muscular tremors for weeks. The man had still not fully recovered after a year. Two volunteers who sat, unclothed, for four days in a heated steel chamber with DDT-coated walls experienced the same, in addition to burning eyes, a feeling of "mental incompetence," and anemia. A group of war prisoners who mistook DDT for flour and baked with it vomited and grew numb, weak, and excited all at once; the men who ate the most couldn't stand on their own, their strength lost for months. A forty-six-year-old farmhand who chewed tobacco on which DDT had spilled developed nausea, vomiting, anxiety, stiffness, and a sore throat. A one-year-old vomited, tremored, and died after drinking an ounce of 5 percent DDT in kerosene.

Forty-six case reports in, Biskind strongly suspected that his own patients' symptoms might be traceable to DDT. Back at his office, he began to ask his patients whether they used or came in contact with the poison.

Every last one of them did. Many applied it to their mattresses to kill bedbugs. Several had DDT-impregnated wallpaper, advertised (misleadingly) as a way to prevent measles, scarlet fever, and polio. Some said their workplaces sprayed it regularly. Some used dry cleaners that added it to their solutions. Two worked with textiles mothproofed with it. The stories were all different, but the DDT in every one of them was the same. Yet Biskind knew that the lion's share of DDT wasn't used in cities like New York but elsewhere, in agriculture. If there was this much DDT used in New York City, what did its use look like in a farming community? To see for himself, he decided to take a trip to the Midwest, where he had grown up and gone to medical school. He hadn't been back in years.

What he saw in the farm towns he drove through shocked him. Cows and other animals were dusted with DDT "with abandon," he said. He had read about the warnings not to spray livestock, but on the farms he visited the animals were doused with DDT in liquid form, too. The barns were all sprayed or fogged, not lightly but "intensively." Everywhere he went, even the feed was contaminated.

In one town he passed through, he picked up a copy of the *Prairie Farmer* newspaper. He was startled to read that cattle across the country were suffering a new disease, dubbed "X disease." It thickened the skin, led the nose and eyes to run, and caused tremors, weakness, a "loss of condition," and diarrhea. He couldn't believe it—yet he could. He had read that DDT was stored in fat and concentrated in the milk of animals. There had to be a connection between X disease in cattle and Virus X in his patients.

Back home in New York, Biskind decided to try to figure out whether his patients were unwittingly eating DDT, too. He bought butter at two different markets and took them to the city's laboratory of industrial hygiene to have them tested. One of the butters contained 8 ppm DDT; the other contained 13 ppm. A part per million is a tiny, infinitesimal amount, equivalent to a drop of poison in a thousand quarts of water. Even so, he knew these levels were higher than the allowable levels of other insecticides, such as lead. And if DDT was stored in fat, that meant those levels would only go up over time in an animal—or person. "Small wonder then," he thought, "that the signs of the virus X syndrome in man and of X disease in cattle are precisely the known signs of DDT poisoning."

Biskind wrote up a report of his findings, laying out a case for redefining Virus X as poisoning by DDT. He sent the report to a medical journal, whose editors decided that its findings were so important they deserved a preview before the article itself was ready to publish. "Special Notice!" the journal's February 1949 issue announced. "DDT Poisoning a Serious Public Health Hazard." Details of the story, the journal promised, would be published the next month.

•

Biskind's report laid out the evidence as he saw it. The national Virus X epidemic coincided with DDT's widespread use by civilians. The DDT poisoning symptoms described in the medical journals mirrored those he was seeing in his patients. The tissue changes described in cases of fatal DDT poisoning in people were the same as those described in reports on DDT poisoning in lab animals. The notion that DDT was harmless was a "myth," he concluded. "To anyone with even a rudimentary knowledge of toxicology, it exceeds all limits of credibility that a compound lethal for insects, fish, birds, chickens, rats, cattle and monkeys would be nontoxic for human beings."

The problem wasn't just that DDT was more toxic than acknowledged; it was that everyone was exposed to its toxic effects. Americans not living on or near farms didn't know about DDT's extravagant use in agriculture. Only a select group of experts knew that DDT accumulated in body fat. And only this group knew that this chemical property meant DDT was present in meat, milk, and butter. Americans had spent the years since the war unwittingly accumulating DDT in their own bodies. The result, said Biskind, was "a large-scale intoxication of the American public."

Though published in a narrowly read medical journal, Biskind's report caused a stir. It made headlines across the country and front-page news all the way to Sydney, Australia. When journalist and fellow New Yorker Albert Deutsch heard the news, he decided to do his own investigation. Deutsch interviewed dozens of doctors and scientists of all kinds to evaluate Biskind's argument. Biskind, he concluded, was "brilliant," his claims all true. "The great bug killer may turn out to be one of the most devastating biological weapons ever loosed by a people upon themselves," Deutsch wrote. "We are eating poison with our fruit, milk and meat, absorbing it and breathing it, all with a heedlessness arising from a false sense of security."

Deutsch was a respected and crusading medical writer with a formidable reputation for social reform. His much-acclaimed books had drawn attention to the deplorable condition of the mentally ill; his articles had led to a reorganization of the Veterans Administration medical program. The nation's top health experts had long praised his work. So when he wrote that the "mounting evidence" of slow DDT poisoning—especially of children—had "so impressed" him, it had an impact. He called for DDT's use to be "drastically curbed" while a presidential or congressional investigation got underway.

As it turned out, related investigations were already in progress, looking into recent changes in food production. During the war, basic food items had been in short supply; milk, butter, eggs, and meat were often rationed, their prices held fixed. After the war, production swelled, and supplies rebounded, but it was just coming to light that wartime shortages had left a lasting mark on how food was made. Synthetic, lab-based fats had been used in place of animal and vegetable fats in processed foods, and a Senate committee was investigating whether their use should be continued. They were cheap but were never rigorously tested. Bleaching and softening chemicals had been added to flour to make bread white and keep it "fresh," and the FDA was just beginning to review whether those were safe, too. The agency was also reviewing—still—the safety of DDT.

The FDA had actually never stopped studying the pesticide. When DDT was released for public sale, the agency was required by the 1938 Food, Drug, and Cosmetic Act to set federal tolerances for the pesticide. Tolerances were first set decades earlier by the Bureau of Chemistry, a precursor agency to the USDA, to limit the arsenic, lead, copper, and other insecticide residues remaining in marketed foods. Under the 1938 law, the FDA, which was part of the USDA at the time, was given sole responsibility for setting tolerances, based on its own research.

In order to set tolerances, FDA pharmacologists ran tests on lab animals to figure out the dose that caused no detectable harm. Then it extrapolated that dose to an equivalent dose for humans (to account for their larger body size and greater daily food intake), cushioned that dose with a so-called safety factor, and arrived at a level that people could likely "tolerate." By law, the agency could set a tolerance only once an insecticide was on the market, and those tolerances were provisional until the agency held hearings on them.

When DDT was first released for sale in 1945, the FDA had set an interim tolerance of 7 ppm for fruit, because that's what it was for lead, and DDT shared one obvious characteristic with lead: it built up in the bodies of lab animals over time. The same observation led the FDA to set a tolerance of zero for milk because so much milk was consumed by babies and children, who were so vulnerable to lead. As yet more studies showed that DDT accumulated in fat, the FDA lowered the tolerance to 5 ppm in any single item of food and kept it at zero in milk.

And then a new farming practice had taken hold, one that Biskind had witnessed on his trip to the Midwest. Dairy farmers and beef cattle raisers began spraying and dipping their livestock in DDT to kill flies. Cattle were notoriously plagued with flies: horn flies, stable flies, horseflies, and more. In summer months in the South, a single cow could harbor thousands of flies. Following a tip from Geigy scientists up in New Jersey, agricultural scientists had shown that fly-free cows gave more milk and put on more weight. A lot more: up to 27 more pounds of weight per season and as much as 100 more pounds of milk per month. Cows that no longer had to swat and twitch, it turned out, put that saved energy into growth. Farmers couldn't say no as county agents and pesticide makers urged them on. "KILL FLIES!" commanded ads in the farm papers. "GET MORE MILK! BEEF!"

There was just one problem: DDT sprayed on cows ended up *in* their milk and beef. Scientists in FDA's pharmacology division knew it would. During the war, they quickly realized that DDT accumulated in animals' fat. Then, in early 1945, they had noticed that after one of their research dogs ate DDT, the pesticide appeared in her milk. A pair of drug-company scientists in Ohio then found DDT in goat milk. Curious to see how much it contained, they fed it to rats and a kitten. The animals died. They made butter from the milk, and the butter killed flies. They sent a report to the journal *Science*, and the farm papers reported it too.

Most farmers sprayed their cows anyway because no one told them not to—until the FDA stepped in. The agency had followed up its tests on lab animals with analyses of milk on the market. Sure enough, it contained DDT. Its pharmacologists set a new study in motion to see whether they could also detect DDT in human fat and milk. But it would take time for them to collect samples, and in light of Biskind's reports, it seemed urgent for the agency to share what it knew. FDA chief Paul B. Dunbar called a press conference. The nation's milk supply was contaminated with DDT, a "cumulative" and "insidious" poison, his officials said. The FDA was going to start spot-checking milk in interstate shipments—the only milk it had any jurisdiction over—and seize those shipments with any trace of DDT.

The FDA announcement caught the USDA and the National Agricultural Chemicals Association (NACA), the industry group representing pesticide makers, off guard. Quickly, they put out statements asking farmers to switch to other chemicals. It didn't help. The public panicked. Milk consumption plummeted. In Atlanta, Lester Petrie clipped the news from the newspaper and slipped it into his files.

A month later, in early May, Wisconsin Republican representative Frank Keefe gave an impassioned speech on the House floor. The public was "unwittingly consuming poisonous chemicals being sold throughout the nation as food," he said. It was up to Congress to put a stop to it. Longtime House Speaker and Democrat Sam Rayburn obliged. He asked New York representative James J. Delaney, a Democrat from Queens, to chair a seven-person committee to investigate the toxicity of the nation's food supply. Before long, Doffie Colson read all about it in the papers.

Chapter 7

POISONED IN OUR OWN HOMES

Lester Petrie, the head of the Georgia Division of Industrial Hygiene, had been sure DDT was safe. Now he wondered whether he ought to create a warning for it, as he had done for TEPP. He reached out again to FDA pharmacology chief Arnold Lehman, who had so helpfully reviewed his TEPP warning. "A few days ago," he wrote, "I noticed a news item in the paper cautioning against the spraying of barns with DDT. I would appreciate very much your sending me all of the new information you may have available concerning the toxicity of DDT."

This time, Lehman didn't reply. Instead, Petrie received a form letter from the FDA's Division of State Cooperation. "I regret that we are unable to send you literature dealing in a detailed way with the toxicity of DDT," it read. Enclosed was a list of articles on DDT poisoning published in medical journals, and an official statement issued by the FDA, PHS, USDA, army, navy, and Pan American Sanitary Bureau, dated two days after the FDA's announcement. Recent news had "misled and alarmed the public concerning the hazards of DDT," the statement read. "DDT is a very valuable insecticide which has contributed materially to the general welfare of the world." It was a "poison" like any other insecticide, but any harm it seemed to cause was a result of "errors" in its use.

Petrie could see what was happening: a feud among federal agencies in the wake of the FDA's milk announcement. He could only guess that Lehman had been silenced. There was surely more to know about DDT, he thought; he just didn't know what. But he knew what was expected of him. When, shortly after, county health commissioner W. D. Lundquist asked him for information on DDT's toxicity, Petrie sent him the federal statement in reply. "I trust this will give you the information which you desire," he wrote.

But Petrie was now more conflicted than ever over the matter of DDT's toxicity. So he continued his quest for more information. He wrote to the CDC, asking for reports on deaths related to TEPP, parathion, and DDT. A scientist from the agency's new Toxicology Branch said that because their work was "just getting underway," the branch had nothing to share. In the meantime, the branch's scientists would be grateful if Petrie could share with them any cases he knew about.

Petrie wrote to the USDA next because he knew it was responsible for ensuring that pesticides were accurately labeled. An agency marketing committee chair sent him some "proposed" label statements: DDT "is toxic," harmful if swallowed, and shouldn't be inhaled, allowed to contact skin or food, or kept within reach of children. Keep in mind, said the chair, that the statements weren't official. But they were all Petrie had, so he copied them into the warning he was drafting on DDT.

In the meantime, his staff had collected information from the Manufacturing Chemists' Association (MCA) and the National Agricultural Chemicals Association. The MCA had published a set of suggested warning labels for economic poisons. Those with TEPP should bear the word "Poison," depict the skull and crossbones, and read "DANGER! EXTREMELY HAZARDOUS IF SWALLOWED, INHALED OR ABSORBED THROUGH SKIN OR EYES." Anything with benzene hexachloride in it should read "WARNING! HARMFUL VAPOR AND DUST." Products containing DDT should be labeled with one of five warnings, depending on whether they were dusts or sprays and what solvent they contained. All of them started with the word "Caution!" The information, Petrie thought, would have been helpful to have years before.

And then Petrie got a call from a public health officer in Bulloch County. A ten-year-old boy in Statesboro had found a whiskey bottle resting in the crotch of a tree, taken a swig, and handed it to his cousin. Before the cousin could take a sip, the boy was foaming at the mouth. In minutes, he was dead. The health officer had the coroner extract the boy's stomach contents, and he sent them to Petrie along with the bottle.

Petrie felt certain they contained TEPP. He also knew that no one, not even the most advanced forensics lab in the state, had the means to identify it. He forwarded details of the case to the Toxicology Branch at the CDC. Then he wrote to the list of manufacturers his staff had compiled, asking them to add tracers to their products so state health officers like him could figure out which products were causing illnesses and deaths. He received one polite reply, from the medical director at Monsanto, which made parathion and DDT. The company would not be able to do this, he said, because it would cost too much.

The economic poisons problem was growing thornier by the day. But Petrie believed that he had enough information to issue warnings on five of the new insecticides: parathion, toxaphene, benzene hexachloride, DDT, and TEPP. He compiled them into a booklet titled *Economic Poisons* and mailed it out to health offices, county agents, and newspapers across the state. He attached a statement of his own, in which he tried to sound serious but not too alarming; he knew it was his job to promote safety, not panic. "Even slight exposure to some of these chemicals may bring discomfort and illness. Even one drop of TEPP concentrate in the eye could cause death," he wrote. "If you use any of them, or if your neighbor uses them, be sure that you understand about them."

Newspapers all across Georgia ran the message just as he wrote it.

•

The summer of 1949 was a bad one for Colson and her family. Her sister was diagnosed with copper poisoning, chronic arsenic poisoning, and partial deafness in both ears. Plyler had saved up to see

F. Levering Neely, a Johns Hopkins–trained doctor who worked out of the grand Medical Arts Building in Atlanta, and he had attributed the deafness to the poisoning and told her to avoid all sprays. Colson wrote to their neighboring landowners to ask them to switch to ground spraying on her sister's behalf. They refused. And then one morning, a plane loaded with parathion flew low over her property, coating her house, flower yard, honeybees, vegetable gardens, and goat and cow pastures. It circled overhead six times, barely clearing their electrical wires and the tops of their pecan trees, before spraying the neighboring cropland. "There was no way to escape the poison," she said. "Not a thing was missed." Crop-dusting pilots could be showy daredevils, but this felt hostile. Colson fell so sick that she spent the next two weeks in bed.

Then a plane for the state came through, spraying their properties with DDT to kill white-fringed beetles. This time, Colson had enough warning to get out of the house beforehand. But not long afterward, she spotted insects she wasn't familiar with on some weeds. They spread to her flowers and then feasted on her roses, their numbers exploding. When she asked a state agricultural agent what to do, he told her not to worry; they were rose beetles, he said, and they were everywhere. Colson had read enough to know what had happened: the DDT spraying had killed the rose beetles' natural predators. And now they needed to figure out how to get rid of the rose beetles without doing more harm.

Then she received a packet from Lundquist, containing the FDA's statement on DDT and an article he had clipped from the Georgia Health Department's newsletter. When she finished reading it, she was sure Lundquist hadn't read it himself, so she mailed it back with a note. The article included a statement from a top USDA scientist, Fred Bishopp, saying that DDT should be used carefully, to kill only harmful insects, and only in the amounts necessary, in order to prevent chronic illness from developing in people. "Please read this paragraph carefully again," she wrote. "I am sure if everyone would pay careful attention to this bit of advice by Dr. Bishopp there would be far less reason for complaint."

•

Five years had passed since Colson had first sought help for the health problems plaguing her and her family, and in that time their troubles had only compounded. Her husband, Henry, still had his job as a blacksmith, but losses on the farm meant that their total income had steadily declined while their doctor, pharmacy, and hospital bills added up. Plyler's farm income had declined even more because her health was so poor she could hardly work. They were all overjoyed when Plyler's daughter Betty was named high school valedictorian and offered several college scholarships. But she'd had to turn them down to take a job to support the family.

Colson had really hoped to start a health movement, but she couldn't find the time or muster the energy, especially when each new round of spraying sapped her strength and health. Plyler, though, had a way of seeing their problem as part of a larger pattern. The problem, as she saw it, was political, which meant that a movement was exactly what they needed. She sketched a plan: letters to political representatives and power brokers, articles in national publications, petitions from small farmers like them, suffering in the same ways that they were. In August 1950, from an Atlanta hospital that Neely had just admitted her to for treatment, Plyler started their movement with a letter to Georgia governor Herman Talmadge. "How or where can we have a chance to live a happy, normal, healthy life in our own home on a small farm? Do we have to suffer the rest of our life and be poisoned to death in our own home?" she asked. "Is Georgia drifting into farming practices that make it impossible for families to own and live on small farms?"

Talmadge, up for reelection that year, quickly replied. "I am sincerely sorry to learn that your health has been so poor," he wrote. Her case was already known to T. F. Sellers, the head of the State Department of Public Health, he assured her. "I am sure he will give this matter a thorough investigation."

But Plyler already knew they couldn't count on the health department. It had been nearly three years since Colson had sent the department her soil, and it still hadn't shared any results. It was time, she told Colson, for them to resume their own investigations and turn their findings into fodder for their movement.

Back home in Claxton that September, Plyler went through town asking anyone she could whether they had been poisoned or sickened by insecticides. Everyone had a story to share. Farmers told her of their poisoned livestock; some told of poisoned workers. "I might be the one to get sick next time," one told her. A bus driver told her of a boy who had died after eating berries grown near a sprayed field. Two other farmers told her insecticides made them ill, and two merchants who sold the chemicals said the same. A vet who had served in the Pacific said that poisons "sure hurt" him and that he could no longer stand to breathe them. She listened to Governor Talmadge on the radio and heard him extol Georgia's "little home owners" as the source of the state's strength. "These little home owners will soon be extinguished if they and all their possessions are covered with insecticides beginning in March and continuing into October every year," she wrote in reply.

The insecticide issue, as Plyler now saw it, was a matter of equal treatment under the law. She and her family had willingly sacrificed for their country, but more and more was taken from them with each passing year. Worse, she believed that the rights of others, especially those with more money, were protected while their own rights were denied.

When Plyler heard that their congressman, Prince H. Preston, was interested in the army camp built on their family's old land during the war, she wrote to him of her troubles. When she read a Channing Cope column warning readers to "be very careful" with sprays, she wrote to him, too. When the pastor of Atlanta's East Point Avenue Methodist Church visited a hospital room she was in, she shared her story again. The pastor, moved, wrote to Governor Talmadge about the family's plight. "I feel much concerned for them," he told the governor. "I do hope that in some way the State Health Department may be able to rectify future situations like this."

Talmadge, Preston, Cope, and both Georgia senators all pressed the health department to investigate. The department now had no choice, and the responsibility fell to Petrie, who now knew all too well how difficult it was to say anything definitive about the harms of economic poisons.

Petrie's staff had no experience testing people for DDT exposure; few had. He had confirmed Plyler's arsenic poisoning diagnosis with Neely, though, so he decided to make that a focus of the investigation instead. He drafted a survey of questions and enlisted the help of Lundquist and Richard Fetz, a wisecracking and good-humored hygienist on his staff.

Lundquist and Fetz arrived in Claxton on a warm day in October 1950, with Petrie's surveys, sampling equipment, and containers for collecting well water and urine. The families on Highway 280 listed their symptoms on the survey's first page, along with the insecticides they used on their farms. But when Colson got to the second page, she felt her blood boil. The survey asked for true or false answers to a series of statements:

There are records of DDT being harmful to humans.

There are records of DDT being found harmful to warm blooded animals and fowl.

In Georgia at the present time DDT is being used in and around dairy barns and dairy cows.

"Won't answer," she told Lundquist.

He insisted that she try. The questions were "not controversial," he said.

But she couldn't bring herself to do it. She knew the statements were all true, and she knew the health department would use that knowledge against her. It was a trap set out to discredit her and her symptoms. "Won't answer," she wrote, halfway down the page.

Her refusal infuriated Lundquist. Colson was "very rude," he complained to Petrie later. Her "poor logic" and "lack of respect" were insufferable. Separately, he wrote up his official conclusions. "There is a good possibility that some of the symptoms claimed by Mrs. Colson and Mrs. Plyler particularly could be due to the insecticide containing arsenic," he wrote. "But I am still unable to convince myself that D.D.T. poisoning can exist as claimed."

Colson and Plyler knew the investigation would find no proof of harm. They spent the next two weeks urging other families on their two-mile stretch of highway to join them in drafting a petition to Sellers, the head of the Georgia health department. Writing "on behalf of all small land and home owners in Georgia," they recalled a time when all insecticides were applied by "a flour sack held over plants" or "an implement mule drawn," settling only where applied. Now farm poisons drifted far and wide. As a result, the poisons were in their eggs, milk, meat, and butter, the water they washed their clothes in, the air they breathed, and the vegetables they grew in their soil. And even the manufacturers warned against such contact. "We the undersigned small land owners," they wrote, "ask you to use the authority vested in the Georgia Dept. of Public Health toward protecting our lives by restricting the uses of, and methods of applying, insecticides in residential areas."

Petrie was defensive when Sellers sent him the petition. Their "previous complaints were originally about DDT," he said, and then Plyler's doctor had confirmed he should test for arsenic. Now they were complaining about all insecticides. He promised Sellers that he was investigating to the best of his ability, even as it meant keeping pace with an increasingly ill-defined problem.

Later that month, the health department sent the families the results. One of Plyler's three wells, the one closest to their house, was contaminated with arsenic. Fetz advised them to abandon it. Some of the Claxton residents had traces of arsenic in their urine, but even the highest level, in a young woman in another family on Highway 280, was still below levels that industrial hygienists considered normal. The findings solved nothing.

Colson and Plyler took their movement to the next level. Plyler wrote to an editor at a "prominent publication." Colson wrote to the American Medical Association, which had announced that it would be looking into DDT as a food poison after the milk scare. She wrote to a congressman in New York she had read about in the paper, James J. Delaney, who was investigating the use of chemicals, including insecticides, in food production. She had no way of knowing just how momentous—or protracted—Delaney's investigations would be; it was just one more thing she thought to do. Then she notified Governor Talmadge, successfully reelected, of her actions. Again he quickly replied. The health department, he said, would "further" its investigation right away.

•

In March 1951, Fetz, Lundquist, another hygienist from Petrie's staff, and a county agricultural agent made plans to travel back to Claxton. It was warm and breezy when they arrived in April. "While you are riding around I hope you get to see insecticide dusting going on," Plyler told Fetz.

He did. Very quickly this time, things changed. At Fetz's request, one big landowner agreed to switch from dust to spray so that his insecticides wouldn't drift as far. Another agreed to have his farmhands load their sprays farther from Plyler's driveway and field. "We were very pleased," Plyler told Fetz as they left.

But a month later, it was as if nothing had changed. On a windy morning, one of Plyler's daughters rushed out to milk before school. Although she could see a crop duster setting out, she milked anyway; it was her chore. She left the milk in the kitchen and ran to catch the bus. Plyler, not knowing, poured a glass. Her mouth and throat burned and swelled. A few weeks after that, Colson got caught in a poison drift that landed her in the hospital. She was released with a deep, unyielding cough. She asked the county agent helping Fetz and Lundquist what was sprayed, and he came right back with an answer: one insecticide containing copper and arsenic, and another containing chlordane and DDT.

Chlordane was new to Colson. She sent a note to the county health office asking for more information about it. She didn't know that the office forwarded her request to Fetz in Atlanta. "We have some information here on chlordane," Fetz replied. "But it is not exactly the type of data that she should get her hands on."

He offered to answer Colson himself and told her to write to a man named Wayland J. Hayes at the CDC's toxicology branch. Fetz then told his colleague in the county health office that although he'd been hoping to stop in Claxton on his way to Savannah that winter, the latest request from Colson left him thinking twice. Colson always seemed to have her "guns loaded," he cracked, so "we may go without fruit cake this year."

•

Without bees, chicks, or health, life in Claxton had become insupportable for Colson, her sister, and their families. There was also less and less holding them there. Colson's father, who had his own Claxton farm, died in early 1951. Within the year, her older daughter married a man from west of Claxton and moved away. Plyler, done farming, talked of moving to Savannah. Colson and her husband decided to pack up their own farmhouse and move to Atlanta with their younger daughter. They rented an apartment on wide and bustling Ponce de Leon Avenue. Henry, still muscular in his sixties, planted a vegetable garden out back. Colson, in her early fifties, enrolled in a program to become a certified practical nurse. Her farming days, and her movement, were over.

She let go of something else, too. She had always loved her country. She still did. But it had changed, she believed. She had lost faith in its new postwar government. It was no longer an institution she trusted, nor did she believe it held her and her family's interests at its core.

Down in Claxton, the spraying continued. Evans County announced a new program to treat homes and properties with a combination of DDT and chlordane in order to reduce malaria and dysentery and make people "more comfortable." The poisons were paired to outwit DDT-resistant flies. White-fringed beetle spraying continued. TEPP stayed on the market. And the farmers who owned the fields of tomatoes, peanuts, and tobacco rented out the land on which Colson's farmhouse sat. Every year from March to October, planes passed over it to spray.

Chapter 8

MEDICAL STANDING

Forty-nine-year-old James J. Delaney was a little-known Democrat from Astoria, Queens, who had just won his House seat back. An Irish Catholic with a law degree from St. John's and time in the district attorney's office, he was easily elected to the House of Representatives in 1944 but ousted in a Republican sweep in 1946. Two years later, though, he was smoothly reelected in a citywide shift back to Democratic party lines.

His first year back was quiet. The war-rattled economy was back on its feet: price-control battles were over, inflation had been avoided, and long lines no longer snaked out of every butcher in Astoria. He sponsored legislation to credit postal workers for time served in the war and a bill to restore citizenship to Queens residents with Italian ties. He earned passing praise for breaking up a fistfight between two representatives on the House floor. Political reporters called him "able." Buzz spread about his mayoral potential. But not much else had distinguished him when Speaker Sam Rayburn tapped him for the role that would define his career.

Delaney set to work assembling his food-safety committee with characteristic diligence, taking no sides and promising to be thorough and fair. He defined his committee's charge carefully and wrote to dozens of food and agriculture companies to enlist their cooperation. His committee, he wrote, was conducting "a study of the effects of the introduction of chemicals into food products, the effect of pesticides on foods, and the effect of chemical fertilizers on foods." He welcomed their comments, recommendations, and witness suggestions. He invited them to send along their policies, if any, on chemicals in food. And he announced that his hearings would be kept open, "in the interests of the public."

The Delaney hearings, as they would quickly become known, began on September 14, 1950. It was an already tense morning in Washington. Polio was in retreat from its summertime heights, but civil rights cases were moving into the courts and Cold War tensions were ratcheting up as US warships bombed the Korean coast. Delaney, setting all that aside, explained his committee's charge, introduced its members, and called the first witness, Paul B. Dunbar, commissioner of the FDA.

Dunbar, who had set off the national milk scare the year before, was a food chemist who had worked in federal food safety since shortly after the first Food and Drugs Act was enacted in 1906. He told a story, familiar to many in the room, about the law's update in 1938. As it happened, the story also explained his agency's circumspection regarding DDT. The 1906 law, Dunbar explained, was in the process of being revised when, in 1937, the FDA received a call from the American Medical Association about a series of deaths in Tulsa, Oklahoma. The AMA had traced the deaths to a drugstore product called Elixir of Sulfanilamide. The medicine's active ingredient was one of the new bacteria-killing agents known as the sulfa drugs.

To make Elixir of Sulfanilamide, a chemist at Tennessee-based S.E. Massengill Company had dissolved the sulfa drug in diethylene glycol, a sweet, odorless liquid. There was just one problem: until then, diethylene glycol had only ever been used as an antifreeze. The resulting elixir was a lethal poison that killed more than a hundred customers. At the time, Dunbar said, nothing required companies to test drugs before selling them, and the FDA lacked the authority to do anything about a potentially harmful drug until the harm had been proven. But the deaths were such a scare that the 1906 law was swiftly rewritten to require all drugs to be tested before being sold. Yet there was a serious omission in the update, said

Dunbar: the revised law had no comparable protections for food. Now, a decade later, more than 840 chemicals were used or proposed for use in foods, and no part of the law required that they be tested for safety first.

Dunbar didn't even mention DDT, but as he neared the end of his testimony, the committee pressed him about it, asking about levels in the current milk supply. Present surveys, said Dunbar, were "reassuring." They asked him whether his agency had any control over DDT and other insecticides' use. "Not directly," he replied.

Two of Delaney's committee members were physicians: Erland Harold Hedrick, of West Virginia, and Arthur L. Miller, of Nebraska. Hedrick asked about the "terrific" headaches his "lady" patients reported after using DDT, and Dunbar realized he'd better take a moment to share his full opinion on the insecticide. DDT was "one of the sensational developments" of the war, he granted, but it was also "poisonous." If not for his agency, whose pharmacologists had found DDT concentrating in fat and milk, and whose agents had called for a halt to cattle spraying, he said, the problem "might have developed into quite large proportions."

But the real problem, Dunbar added, was bigger than DDT. Because of the Federal Insecticide, Fungicide, and Rodenticide Act (FIFRA), passed in 1947, some sixteen thousand insecticide products had been registered for use in the last few years. Under that law, pesticide makers had to register and label their products, but they did not have to study their toxicity or where they ended up once applied. The FDA set tolerances on pesticides, but nothing stopped manufacturers from marketing insecticides before tolerances were set. Some insecticides, warned Dunbar, might even break down into more-toxic compounds after application. "There were even insecticides put out," he said, "for which no chemical method of identification or analysis is known." DDT was just one of them.

Dunbar was trying to make a point about DDT: it was just one representative of an entirely new landscape of insecticides. But that wasn't the point committee members took away. On that first morning of the hearings, DDT became their benchmark for understanding the promises and challenges of the new postwar insecticides. To

Delaney, that meant he should call more witnesses to testify about DDT. But this focus on DDT would end up creating a vulnerability that insecticide manufacturers would later use to their own good.

•

Morton Biskind had never had as prominent a podium as Delaney granted him that fall. He had worked for the AMA's councils on pharmacy and chemistry right through the Elixir of Sulfanilamide crisis—but that episode earned only its manufacturer any fame. From there, he was appointed head of Beth Israel Hospital's endocrine lab and clinic, where he earned some minor attention for research on how the B vitamins regulated sex hormones. But he had relinquished that post not long ago and moved with his wife to Westport, Connecticut, commuting by train to New York to see the patients in his private practice. Their persistent waves of symptoms continued to convince him that he had just made one of the most important observations of his career.

Seated before Delaney's committee, Biskind began to share stories about his patients. He told them about the ailing exterminator who had sought relief from various doctors for two and a half years, eventually having his skull opened in search of a brain tumor. When he switched jobs and removed all pesticides from his environment, his health improved in a week. He described the woman who had spent more than two years confined to bed, told by multiple physicians that her symptoms were "fantastic," only to recover within days after removing all insecticides from her home and diet. A man sick for eight months experienced relief when he stopped eating slabs of butter with each meal. A third of Biskind's patients, he estimated, had DDT-poisoning symptoms. His advice to them was to take down their DDT-treated wallpaper; wax over DDT-treated walls; dry-clean the DDT from their clothes, drapes, and carpets; consume peeled fruits; and avoid butter and beef. Without fail, they got better.

But most doctors, Biskind told the committee, were unaware of DDT's effects. They were difficult to identify because the signs of DDT poisoning were so easily attributed to other causes. The muscle weakness mimicked polio, the liver damage seemed to point to hepatitis, and the vascular degeneration pointed to heart disease. Depression and other mental conditions could be explained by the poisoning's unyielding symptoms, but instead they were chalked up to mental dysfunction. Some portion of Virus X cases might be attributable to some yet-to-be discovered viral agent, he conceded, but in his own practice, he thought, many were undoubtedly caused by DDT.

Biskind went on to describe the lab tests he had conducted. Butter had high levels of DDT, as did beef. He tested milk from a dairy in Connecticut that had never used DDT, and it still contained 0.5 ppm DDT; the cream contained 4 ppm. One brand of cigarettes contained 4 ppm. The breast milk of one of his patients contained 116 ppm. When she breast-fed her newborn, he said, her symptoms subsided as the chemical flowed out of her body and into her baby. He tested breast milk from five other women; all of it contained DDT. As he spoke, he felt his indignation rise. Thanks to "some of the fanciest double-talk ever perpetrated," he said, the public was experiencing a massive epidemic of misdiagnosis by doctors unaware of the facts about DDT. "Is there any apologist for DDT who maintains that this poison is a proper ingredient of mother's milk?"

Senator Miller interrupted. He was worried, he said, that Biskind's testimony was "sensational." He asked if Biskind was aware of any "large group" of "scientific men" who had reached the same conclusions.

No, said Biskind. Just a few. He shared their names with the committee.

Then Senator Hedrick, the committee's other physician, interjected. Was it possible that all of these patients were merely hypersensitive to DDT?

They most surely are, Biskind said. But that didn't make DDT harmless.

Hedrick, however, suspected Biskind might be speaking beyond the limits of his expertise. Can you draw DDT's chemical structure? he asked.

"Certainly," Biskind said. He described the compound's two benzene rings with their respective chlorine atoms, attached in the middle by an ethane group with its own three chlorine atoms attached.

Miller still bristled. DDT was gold to the dairy farmers in his state, who were prospering, and he knew of no poisoning epidemics. He addressed Delaney this time. Biskind's testimony was "doubtful," he said. He asked that the doctor's statement be expunged from the record.

Delaney resisted. Biskind had impeccable credentials, he said. And they had all heard testimony from Arnold Lehman, the FDA pharmacologist, who had told them that the artificial sweetener dulcin was on the market for fifty years before scientists learned that it was toxic to rats. There was a "constant struggle" to ascertain the safety of chemicals used in food, Delaney said. Speaking of Biskind, he said, "I think this doctor has done yeoman service. He is to be complimented."

Miller wasn't satisfied. "So whom are we to accept here? A small segment, one individual?" he asked. "Or are we going to accept the larger group of scientific men and Government agencies who, too, have made determinations?"

"I think pioneers have always been in the minority," said Delaney. "Not only in scientific study, but I think in most other things. And it takes a man such as this doctor to bring these facts to the attention of others."

Reluctantly, Miller retreated. He would accept Biskind's statement, he said. "But if we do accept it as being true, then the government agencies are going to be severely reprimanded from this day on for permitting or giving the green light to the use of DDT."

Delaney set the comment aside. Biskind's testimony, he said, "is as valuable as any testimony that we have had before this committee. And I for one am very interested and eager to hear the rest of it." He turned to Biskind: "I wish you would continue, doctor."

Biskind did not want to miss his chance to make three more points. He described studies by FDA scientists that showed that when a person's fat was "mobilized"—either through dieting, starvation, or some other cause of weight loss—acute DDT poisoning occurred as the cumulative DDT stored in the fat was metabolized all at once. He also wanted the committee to know that claims of DDT's safety were based on a select group of "young, active, healthy male adults"—army personnel—whereas now the whole population was exposed. And not just to DDT: in a single day's diet, Americans were now consuming DDT, benzene hexachloride, chlordane, chlorinated camphene, methoxychlor, parathion, and other poisons, in addition to the lead- and arsenic-based compounds still in use. "How many simultaneous insults can the human body take?" he asked.

At the end of his statement, the Delaney committee's chief counsel, longtime government solicitor Vincent A. Kleinfeld, asked Biskind to read aloud the federal statement of DDT's safety drawn up in the wake of the milk scare and signed by the army, navy, USDA, FDA, and Pan American Sanitary Bureau. The signatories were, Biskind commented after reading, "quite a formidable group."

"Doctor," said Miller, "it seems that your views are considerably at variance with the views of these various—"

"Not entirely," Biskind interrupted. "If ten years ago the same agencies were asked whether Agenized flour was safe, they would have insisted that it was." The chemically bleached flour was marketed for years before scientists showed that it was neurotoxic to dogs.

Delaney agreed. "There is nothing definite in science, is there?" he said. "Doctor, we want to thank you for your well-prepared statement. It is quite a statement, I might say."

Hedrick didn't say it that day, but he agreed. Biskind's testimony, he confessed later, "scared us nearly to death on the use of DDT."

•

The members of Delaney's committee reached the end of 1950 convinced they needed to learn more. They had heard from more than a dozen witnesses after Biskind: organic farmers, food chemists, soil scientists, and Mrs. Harvey Washington Wiley, widow of the man who had crusaded for the 1906 food law, among them. The picture the witnesses painted was "alarming," Delaney's committee told the press and pointed to the need for a new law to control chemicals in foods. And yet there was still more to understand. With Congress's approval, Delaney's committee extended the hearings another year, hoping to bring in more witnesses from food companies.

But the companies hung back. "They have avoided us like the plague," Kleinfeld complained. Government and university scientists, on the other hand, were eager—especially if, like Wayland J. Hayes, they were hoping for a chance to refute testimonies already given.

Hayes was the little-known director of the CDC's brand-new toxicology division in Savannah, Georgia, where the CDC's early DDT tests were conducted. It was the unit that had asked Petrie to share his poisoning cases as it got off the ground two years before. Hayes was appointed its temporary director at age twenty-nine—and then kept on. He was ambitious, quick thinking, and technically qualified, with a master's degree, an MD, and a PhD in zoology and physiological chemistry. His supervisor had been quickly convinced that a more experienced scientist couldn't do a better job.

The Toxicology Section itself was created to fill a gap in federal scientific capacity. The FDA studied economic poisons' effects on animals, but aside from a few wartime tests at the National Institutes of Health (NIH), conducted by industrial hygienist Paul A. Neal, no one was studying their effects on people. During the war, Neal's scientists had put two men in small chambers filled with aerosolized DDT for an hour a day, six days in a row. They followed that up with a similar study in which the volunteers were stripped to the waist. A year later, they fed one of the volunteers 475 milligrams of DDT dissolved in olive oil. Six months after that, they fed him a second, larger dose. They also carefully examined men, like the recruit from Fort Dix, who had worked as sprayers. In none of the men did they note anything of concern. Hayes's job was to pick up where Neal left off.

Toxicology was still a developing discipline when Hayes took over the lab at the CDC. That gave him a choice to make: he could study the toxicity of economic poisons as a pharmacologist would or as an industrial hygienist would. The FDA's scientists approached toxicity as a pharmacological question, administering progressively larger amounts of a poison to various lab animals and documenting the full range of the chemical's effects on their cells and tissues, no matter how minor, marked, or delayed. Industrial hygienists, such as Neal, approached toxicity differently. Theirs was the approach that Hayes took.

The field had emerged in the late nineteenth century to ensure that industrial or factory workers were safe and healthy enough to stay productive, even as their work put them in daily contact with fumes, dusts, caustic brews, noises, heat, and other hazards. Hygienists focused intently on a hazard's acute (short-term) effects on the human body. If something made a worker sick today or unable to work tomorrow, it needed to be addressed. To make the connection between a hazard, or "exposure," and its effects, hygienists adapted an idea from "germ theory," then ascendant in medicine. Germ theory was the paradigm that connected specific microbes to specific diseases, such as cholera or tuberculosis. Germ theorists showed that a germ always caused the same, predictable, visible symptoms; industrial hygienists came to assume that the same was true for chemicals. Only those visible symptoms known to be associated with a chemical—from studies of the highest "tolerated" doses in animals and people—should be attributed to it. Any other symptom, industrial hygienists believed, must have some other cause. And like germ theorists, industrial hygienists assumed that the only thing that made a poison hazardous was the size of the dose.

When Hayes appeared before Delaney's committee on an April morning in 1951, Neal sat beside him. Hayes spoke first, reading their jointly prepared statement, which was focused entirely on DDT. They offered it as a "comment on the testimony presented by Dr. Morton S. Biskind."

Hayes described Biskind as an unscientific man who believed flawed reports, carried out no careful studies of his own, and took his patients at their word. Biskind, complained Hayes, had not verified his patients' descriptions of their DDT exposures or documented exactly how much DDT they had been exposed to. He hadn't compared his patients to any sort of control group. He claimed his patients were experiencing symptoms of chronic DDT poisoning, but there were no other "authentic" cases of such in the literature.

Hayes then went through several of the published cases of DDT poisoning Biskind had described, noting that the supposedly poisoned individuals' symptoms could all be explained by some other cause, from DDT's solvents to "life in a war-torn country." Regarding the experiment in which two men were exposed to DDT in a heated chamber, Hayes said their symptoms would have been felt by "anyone who had to sit on a hard bench" for that long. Published cases of DDT poisoning, he said, proved nothing about Biskind's own patients. In fact, he strongly suspected that Biskind's patients were suffering from "a hysterical type of psychoneurosis." It was an argument Hayes would return to repeatedly in the decades to come.

Hayes didn't deny that DDT was toxic or that it was present in people's bodies and in breast milk. Researchers in his own lab, in fact, had collected fat samples from people in Savannah and Washington state's Wenatchee Valley, finding DDT in both. They were also finding high DDT levels in fat samples from people who mixed or applied DDT for a living. But in none of these people, insisted Hayes, was the insecticide causing poisoning.

When Hayes finished reading his statement, Kleinfeld asked if regularly ingesting small amounts of DDT could lead to poisoning. No, Hayes said. But then Neal stepped in. A large enough dose of DDT could cause death in all species, he said. Repeated large doses could cause acute poisoning, he added. And then he passed the buck. The question of small doses, he said, had been left to the FDA, whose research had shown that small doses were stored in fat.

"So . . . when DDT is consumed, and I quote your language, 'in relatively large amounts,' it is a highly toxic substance?" Kleinfeld asked.

"It is a highly toxic substance," Neal answered. "We would like to emphasize that."

"And it is now regarded as more toxic than it was considered to be a number of years ago; is that not correct?" Kleinfeld asked.

"Yes, sir," said Neal. "I think that is a fair statement, especially due to its accumulation in fat and excretion in milk."

"Now, in view of the fact that it seems to be established that even small amounts are stored in the fatty tissues of the body, may it not be that the ingestion of comparatively small amounts of the compound may produce chronic human illness?" Kleinfeld asked.

"We are simply stating that none has been reported," Hayes said.

"Have the eating habits of the population been studied?" asked Kleinfeld.

"We are in process of studying that now," said Hayes.

Hedrick interjected: "Dr. Hayes, do you believe there can possibly be any connection between DDT poisoning and Virus X?"

Several years into its spread, Virus X remained a medical mystery, evading any form of consensus regarding its cause. A top army commander had declared it a result of "bacteriological warfare" by Russian agents. A Texas doctor said that it had turned out to be mere gastroenteritis. The head of California's health department announced that in that state, at least, it was type A influenza infection combined with "epidemic nausea." The conclusion was published in the *Journal of the American Medical Association*. Hayes referenced the publication and then dismissed the idea that the term "Virus X" meant anything. It was a catchall phrase used to describe everything and nothing at once, he told the committee. It "has no medical standing now at all."

Chapter 9

DELANEY'S CLAUSE

Among the many questions James Delaney's committee next struggled to answer was whether the DDT already accumulated in people's fat was safe. They asked every relevant expert who testified, but none would give them the answer they needed. Scientists were accustomed to thinking about safety as a relative and uncertain concept. Government scientists were constrained by the political nature of their positions. So they said what they could, which was often very little. Arnold Lehman, for one, noted that 5 ppm caused liver injury in rats. Wayland Hayes wouldn't answer, referring the committee to the FDA. Edward Knipling—who had first discovered the powers of DDT for the army in Orlando—spoke in circles and then also pointed to the FDA. When Vincent Kleinfeld put the same question to Fred C. Bishopp, assistant chief of the USDA Bureau of Plant Entomology and Quarantine, he too referred the committee to the FDA.

"So the Department of Agriculture and the Food and Drug Administration have different ideas on the possible hazards of DDT?" asked Kleinfeld. "Is that correct?" In his long-winded reply, Bishopp never directly answered the question. But it was crystal clear that he might as well have said yes. Over the next hour of questioning, Miller pieced together a new view of the DDT problem. The USDA and the FDA had overlapping authority over pesticides and completely opposing assessments of DDT. "There will have to be a little surgery done here, a little grafting of departments," Miller said, "if

96

we are going to get any place in this business of cutting down on our bureaucracy and conflicting authority and responsibility."

Bishopp disagreed. His agency, he argued, had an impeccable safety record, with no cases of injury recorded in accordance with its insecticide recommendations. "I wish we had as good a record in the field of drugs," he said.

It was a calculated jab. And with that, what historian Frederick Davis calls "the fundamental fracture" in the regulation of poisons had been laid bare. The USDA was responsible for figuring out whether chemicals were getting into the food supply—and it did. The FDA, which had split off from the Department of Agriculture in 1940, was responsible for determining whether those chemicals posed a problem for human health. Because some of those same chemicals were used in public health efforts, the CDC also bore some responsibility. And with labs in all three institutions adhering to different scientific frameworks, the result was what Davis called a "cloud" of scientific uncertainty and a bureaucratic stalemate.

In the meantime, Miller moved on. He joked that a simple solution to the problem might be a national advertising campaign that told mothers that DDT in milk would make their children bigger.

"We would not recommend it," replied Bishopp.

"I'm not trying to be facetious," said Miller. "I am trying to indicate the power of advertising."

•

Delaney's hearings spanned three years. In the second year, the committee took its show on the road, gathering testimonies in six other cities across the United States. By the end, more than two hundred witnesses had appeared before them. Farmers told stories of pests outwitting sprays. Entomologists told of DDT-resistant flies and aphid explosions, ushered in when DDT killed their predators but not them. Representatives from baby food companies said their customers wanted products made with crops that hadn't been sprayed, but they struggled to find any.

DDT was not the committee's focus, but insecticides generally constituted a significant portion of the testimonies it heard. Witness after witness detailed DDT's harms, called for more regulation, or recommended that it—and other insecticides—be taken off the market. And just as many witnesses said that DDT was, as one expert put it, "innocuous."

DDT had been in use seven years, yet scientists sharply disagreed over whether it was, on balance, hazardous or safe. It was paradoxical that this was the case, said a witness from the University of Cincinnati, for DDT had "been subjected to more scientific investigation than any other organic material." Yet more scientific study seemed only to lead to more disagreement.

In part this was because there was no consensus on how to set the terms for assessing DDT's hazards outside the context of war. It was clearly much less acutely toxic than such poisons as arsenic or TEPP. To scientists who adhered to the tenets of industrial hygiene, as Hayes did, that made it safe because they believed safe doses existed for all poisons. But some people seemed more sensitive to it than others. And a larger question, for some, was what harm DDT might be capable of in the long run, especially because the body stored it over time. California physician Francis Marion Pottenger, who had collaborated with FDA scientists on a study that found DDT present in fat and milk samples from seventy-five people in his town, worried that too few scientists were thinking about this. "If people . . . have from zero to 33 parts per million of DDT in their fat," he told the committee, "what is going to happen after we have been using DDT from 10 to 15 years? That is the serious thing to me."

The split in scientific opinion left Delaney's committee "in a fog as to whether DDT is a safe thing to use," said Miller. "We are at sea when experts disagree."

•

That tension left its mark on the final report Delaney submitted to Congress in the summer of 1952. It walked a careful middle ground. "At this stage of our civilization, there is a genuine need for the use of many chemicals in connection with our food supply," Delaney wrote. But the public also needed protection against mistakes, "irresponsible" actors, and the "premature enthusiasms of chemical manufacturers."

Of 276 unsafe chemicals brought to the committee's attention, Delaney's report singled out a handful, DDT among them. Overall, it called for a new provision in the Federal Food, Drug, and Cosmetic Act, comparable to the new drug section of the law enacted in 1938, requiring manufacturers to submit acute and chronic safety test results to the FDA for any chemical to be used in or on foods.

But half the committee members took issue with Delaney's assessment. Although pesticides were poisons, said Miller and California Republican Gordon McDonough, commercial producers were acting responsibly. It was the "housewife" who failed to realize that DDT, chlordane, selenium, and other poisons needed to be used with caution. In two minority reports, Walt Horan, a Republican fruit grower whose Washington district included the Wenatchee Valley, and Thomas G. Abernethy, a Mississippi Democrat, found more to disagree with. Delaney's report was unnecessarily "alarmist in nature," Abernethy wrote. The FDA's existing authority to set tolerances and USDA's registration process were sufficient to protect the public from any insecticide hazards, the two agreed.

Across the country, a handful of newspapers ran stories on some of the committee's recommendations as they came out. The committee believed that fluoride in drinking water was a matter best left to the states, they reported, and chemical fertilizers needed no federal legislation. But otherwise the press largely ignored the committee's reports. And the question of DDT's toxicity, a headline-grabbing issue in 1950, was no longer news in 1952.

There was an unseen force behind that shift, a public relations campaign bankrolled by the chemical industry. When Delaney had first announced his hearings, members of the Manufacturing Chemists' Association, an organization of top executives from major chemical companies, had met to discuss what the hearings might mean for the industry. They agreed that at the very least, the association should create its own committee to pay close attention to the hearings and propose any actions they might want to take in response. The resulting Chemical in Foods Committee was composed of executives from Dow, DuPont, Monsanto, Merck, Pfizer, American Cyanamid, the Hayden Chemical Co., and Victor Chemical Works. And they all agreed on one thing right away: the hearings were bad publicity. They pooled money to hire a Manhattan public relations firm, Hill & Knowlton, and they invited its top partner, John Hill, to join them at their monthly meetings at the tony Union Club on Park Avenue.

Hill designed a campaign that his firm would later tailor for the tobacco industry. His publicists drafted a "backgrounder" for editors and commentators at national news outlets, giving them the "facts." They put top chemical executives on the radio. When Delaney's committee arrived in a new city, they flooded local reporters with press releases. The campaign pushed out two messages: chemicals were crucial, and chemicals were safe.

While Delaney's hearings progressed, Hill & Knowlton's publicists cultivated relationships with journalists, convincing them to report on how chemicals made foods more nutritious and kept them free of disease-causing microbes. Newspaper editorials warned that if chemicals in food were regulated, jobs would disappear, and food would become scarce and expensive. Columnists dismissed the hearings as unnecessary fuss. "Foods themselves are chemicals," wrote one. "The salt and sugar grandma used in home canning are home chemicals."

While Hill & Knowlton worked the media, MCA's committee prepared draft legislation to present to lawmakers when the time was right. When, in 1953, Miller introduced a bill to require premarket testing for all chemicals in food, the MCA committee was ready with a response.

The industry group came to Miller with a proposal for an ad hoc approach to testing food chemicals, in which companies would notify the FDA of their plans to use a chemical and then wait for the agency to decide whether or not to require safety tests. Miller—bending to please the industry, agricultural interests, and the FDA—modified his bill. In the end, what became known as the Miller Pesticide Amendment to the Food, Drug, and Cosmetic Act required manufacturers to seek a tolerance from the FDA before a product was used or marketed, and it directed the FDA to set tolerances—but only "to the extent necessary." The tolerances should be low enough to protect public health, but not so low that they compromised the nation's need for an "adequate, wholesome, and economical food supply."

The amendment, passed in 1954, was a balancing act. It passed the House with no debate and the Senate with close to none. It "legitimized" insecticide use, according to one critic, by setting legal doses that were considered harmless. It also legitimized the term *pesticide*, defining it as any substance that was an "economic poison" under the Federal Insecticide, Fungicide, and Rodenticide Act (FIFRA). It did nothing to seal the fracture between the USDA and FDA over pesticide regulation, despite Miller's complaints. The USDA retained responsibility for pesticide registration, and the FDA retained responsibility for setting safe levels for the food supply. The task turned out to be a fool's errand for the FDA, with tens of thousands of already registered pesticides in need of review.

Over the next three years, lawmakers introduced several bills to regulate other chemicals in food. The lot of them, all inspired by Delaney's hearings, opened up a policy debate over the extent to which pesticides should be covered by any new legislation to regulate other chemical additives in foods. Delaney, convinced by his own hearings that much stricter regulation was needed, drafted a bill that included a provision barring from the food supply *any* chemicals known to cause cancer in animals or "man."

Delaney's bill was worked over endlessly as he sought support for it on Capitol Hill, but he worked hard to protect the cancer clause. Several times, the clause was struck. Each time, he went personally knocking on doors to work it back in. Ultimately, it made it into law. The Food Additives Amendment, which came to be known as the Delaney Clause, made it illegal for any chemical capable of causing cancer to be used in the food supply.

In sharp contrast to Miller's amendment, the 1958 Delaney Clause called for "no tolerance." It rejected the idea that there was a safe dose of a chemical that caused cancer. It applied to any chemical in food, pesticide or otherwise. The only narrow thing about it was its definition of harm: chemicals that caused cancer would not be allowed in the food supply, but chemicals that caused other health problems could still be used, as long as the dose was safe. That distinction existed, the amendment stated, because some chemicals were necessary or unavoidable, "in particular the pesticide DDT."

•

Hayes, back in Savannah, had just published the first of the studies he had told the committee he was working on, regarding DDT in the diet. The studies, carried out at two prisons in Tallahassee and Atlanta, would become a key part of legal and congressional testimonies he'd be asked to give in the decades ahead—although that wasn't why he was carrying them out. He was simply aiming to show whether or not DDT in the diet did any harm.

Hayes and his colleagues recruited fifty-one men in the prisons to swallow up to thirty-five milligrams of DDT every morning with a bit of milk, for close to two years. They were "free to leave the study at any time," Hayes wrote in his first report on the study. A good number of them did.

Those who stayed stored high levels of DDT in their fat, up to 619 ppm. But Hayes's scientists found no "evidence of injury." A few of the men in the first prison complained of pain, headaches, and other symptoms, but on examination Hayes's team found them healthy. When they checked some of the men again five years later, they found them all in good health. One man had complained extensively of physical symptoms, but Hayes's team concluded that "his complaints obviously were of psychoneurotic origin."

All the while, scientists from Hayes's division and elsewhere in the CDC responded to complaints coming from the orchards of the Wenatchee Valley and the Mississippi Delta, where cotton was sprayed with aldrin, dieldrin, toxaphene, and benzene hexachloride, often with DDT added. In the delta, an area inhabited predominantly by Black tenant farmers, CDC scientists found that those farmers most exposed to insecticides suffered more headaches, cold symptoms, body aches, skin problems, and gastrointestinal complaints. One man, dusted with parathion, suffered nausea, vomiting, and lasting weakness. A man who accidentally swallowed aldrin developed a headache, nausea, burning skin, and jitteriness. The reported symptoms echoed the earliest reports of Virus X. But the CDC scientist who studied the situation dismissed pesticides as the problem. The main issue, he said, was "the widespread use of toxic compounds by a class of inadequately trained and sometimes careless people."

After a series of studies tested various theories on the causes of X disease in cattle—insecticides, oil-based insecticide carriers, vitamin A deficiency, and others—the USDA published its final word on the cause in 1954. In cattle, the agency determined, the condition was poisoning by highly chlorinated naphthalene, a wax used in pellet feed for livestock.

Meanwhile, physicians gave less and less attention to Virus X in people, although people still claimed to have it. Biskind continued to puzzle over it, still convinced that DDT had some role. Eventually, a famous science writer reached out to ask him about it, and he gave her his honest opinion. For a while, they exchanged letters. Eventually, they decided to meet.

Chapter 10

MOSQUITOES

In the first seven months of 1955, Paul F. Russell spent just five days at the home he had just bought on the picturesque inlets north of Boothbay Harbor, Maine. The rest of the time he was overseas: Ceylon, India, Pakistan, Yemen, Italy, Switzerland, England, and Mexico. He returned to Maine for the month of July and then went abroad again, to Turkey, Iraq, Syria, Lebanon, Jordan, and Egypt.

Russell, a circumspect physician with thin lips, a strong brow, and opinions to match, was one of the world's leading malariologists, and he was in high demand. Since the 1920s, he had worked for the International Health Division of the philanthropic Rockefeller Foundation, where he had devoted himself to the study and control of malaria. During World War II, General MacArthur had called on him to mastermind a malaria-control plan for the South Pacific. When the war was over, President Truman gave him the Legion of Merit. Then the brand-new World Health Organization (WHO), created as part of the United Nations after the war, invited him to join its committee of global malaria experts.

So he had always traveled, just not at this pace. He had weekly diarrhea in Venezuela and severe bronchitis in Bombay. His wife came down with fever and intestinal distress in Karachi. In the heart of Ceylon, near the ancient ruins of Sigiriya, he rang in the New Year watching fireworks and listening to the chants of Buddhist priests.

As he traveled through Ceylon (now Sri Lanka), he visited towns where malaria, once a "great scourge," had become rare. He saw entirely new towns in places previously uninhabitable because of the disease. In 1946, just after the war, more than four hundred of every thousand people living in Ceylon had malaria. Now just eleven did. And it was all because of DDT.

Russell went on to tour a DDT plant under construction in Delhi. He attended Pakistan's first-ever malaria conference. Then he boarded a ship for Egypt. When he got to Geneva in the middle of February, he met with other malaria experts at WHO's headquarters to discuss an idea whose time, they all believed, had come: a switch from malaria control to malaria eradication. Malaria control was an ongoing project that involved never-ending effort and resources to treat sick patients and reduce populations of disease-spreading mosquitoes by killing the insects and their larvae and eliminating the places where they bred. Eradication, by contrast, had an end. If a country—or the world—got rid of every last case of malaria, it would no longer have to engage in malaria control. With a global health organization in place, that now seemed possible. The WHO's malaria experts could lead the way.

While Russell had been traveling, bearing the title of consultant to the WHO, the organization's executive board—its body of technical experts—had formalized the plan in a resolution. WHO's director-general, a physician and former health official from Brazil named Marcolino Candau, asked Russell if he would help present the resolution at a meeting of the World Health Assembly, the WHO's decision-making body. The assembly would be convening at the National Autonomous University of Mexico at the end of May.

Russell went home to Maine for a few days in April and then packed up again, this time for Mexico City. When he arrived, he was struck by the main campus's modern buildings, with their expanses of glass and brightly colored murals. The meeting brought together delegates from seventy-six member countries of the WHO. The delegates elected a president, from Mexico, and three vice presidents, from Iraq, Austria, and India. They discussed the need to expand the organization's work on polio and to protect populations from atomic radiation. They voted to admit Sudan as a new member. And then they approved the resolution from Geneva, voting to create and administer a fund to help member nations around the world eradicate malaria. Russell recorded the outcome in the diary he kept of his travels.

"These were history making decisions," he wrote.

•

The WHO, created in 1948, was the first global organization devoted to public health, and it had spent its first years helping war-torn Europe recover. It had also supported a number of demonstration projects to show how coordination and cooperation could reduce intractable diseases and other health problems. In projects in Argentina, Ecuador, South Africa, and a number of other countries, spraying houses with DDT drove malaria rates down to previously unimaginable lows. By the early fifties, the organization's focus had shifted away from Europe and toward health problems in what were just then called Third World countries. And malaria, most international health specialists agreed, was one of the most debilitating diseases in such countries.

The disease was spread by mosquitoes belonging to the genus *Anopheles*. Anophelines, as they were known, spread malaria parasites, or plasmodium species, with their bites. Earlier in his career, Russell had worked to control malaria in India and the Philippines. Mosquito nets protected people from being bitten as they slept. Drugs such as quinine and chloroquine destroyed the parasites in an infected person's bloodstream. Pesticides sprayed over malarial landscapes killed the mosquito vectors and their larvae. But the Malaria Eradication Program, as the WHO's effort was called, did something else entirely, borrowing the approach tested by the CDC's Savannah branch during the war. It involved spraying DDT, but only on the interior walls of people's homes, because so many anophelines liked to bite people indoors and then rest on a nearby surface immediately after.

Indoor residual spraying, as the approach was now being called, was a brand-new "trick," Russell said. If the walls or ceiling nearby were sprayed with DDT, an infectious mosquito would most likely die before it spread its parasites to another person. It seemed a miracle to specialists like Russell that killing the mosquitoes alone, in such a targeted fashion, could yield such dramatic results. But it did.

The idea to make the approach the cornerstone of a global campaign focused on eradication was inspired in part by the WHO's early demonstration projects and in part by earlier Rockefeller-funded campaigns. One such campaign, to eliminate yellow fever from the Americas in the thirties and forties, had targeted mosquitoes with pyrethrum and their larvae with vast outdoor applications of Paris Green, the arsenic-based insecticide from the nineteenth century. The campaign was expensive and cumbersome but fairly effective, driving the disease out of coastal and urban areas and its rates way down. The other campaign had almost entirely eliminated malaria from Sardinia, Italy, in just five years. There, 10,000 tons of DDT were applied to an island of 1.2 million people—for an average of nearly 17 pounds of DDT per person. Targeting the insect vectors where the insects bit seemed more strategic by far.

The WHO eradication program—the organization's first—got underway in 1955, building on its existing demonstration programs and bringing other nations on board with the help of Russell's diplomacy. In participating countries, teams of sprayers—reporting to supervisors who reported to the highest levels of government—went house to house with knapsacks full of DDT they sprayed on ceilings and walls. In parts of Pakistan the spray "men" were schoolboys. In Ruanda-Urundi, the knapsacks were so large that one man wore them while two other men pumped and managed the enormous hose. After the spray teams retreated, another team followed to test residents for infection. If the parasite swam in their blood, they received an antimalaria drug, usually chloroquine. Then the cycle repeated, sometimes every month, sometimes every several months, until malaria slowly faded away.

The approach reflected a fairly dramatic shift in thinking about malaria control from the prewar period. Back then, investing in economic uplift was seen as the best way to combat disease. Now disease was placed first: if malaria could be eradicated, the thinking went, then economic development would follow. And with DDT, health experts hoped, it would follow fast.

But everywhere that Russell went, he had conversations with colleagues and local officials about the program's most vexing problem. Anophelines were already growing resistant, or as some scientists called it at the time, "immune" to DDT. In Bahrain, DDT no longer killed Anopheles *stephensi*. Anopheles *sundaicus* was resistant in parts of North Java. Anopheles *sacharovi* was resistant in sections of Lebanon and Greece. Greece had turned to BHC and dieldrin instead, and now Anopheles *sacharovi* was resistant to those, too.

The substitute insecticides posed their own problems. When Venezuela switched over to dieldrin, the chemical killed large numbers of cats. As the cats died, rat populations exploded. Dieldrin was sickening the country's spray men, too, causing tremors, double vision, and, in the worst cases, seizures. Russell met a doctor who had examined the men. He said he had found signs of "real disturbance." But when a chemist from a dieldrin manufacturer checked on the workers, he blamed the symptoms on their carelessness. The workers had only been poisoned, he said, because they failed to bathe, slept in their work clothes, and refused protective gear because of the heat.

To Russell, the episode was simply proof that DDT was the best tool for the job because it was not just effective but safe. The episode was also proof that eradication had to happen as quickly as possible. If it didn't, the mosquitoes would continue to become immune. In each country he traveled to, Russell told the ministers and officials he met with that the program was one of "great urgency." He didn't always say as much, but it was right there in the WHO's resolution: eradication had to be carried out before the chlorinated hydrocarbons no longer killed anophelines, which were rapidly becoming resistant.

Neither Russell nor his colleagues saw spreading resistance as any reason to hold back. They believed that eradication was feasible. They simply saw it as a race against time.

•

For Americans like Russell, however, it was also a race against the Russians. The malaria-eradication program took place against the backdrop of the Cold War. The Soviet Union and the Eastern Bloc countries had pulled out of the WHO a few years earlier, but when the eradication program commenced, they were just beginning to rejoin. From the very beginning, the contest between the United States and Russia, two of the most powerful WHO members, turned the program into a battle between capitalism and communism. Officials at Rockefeller and at the aid-granting agency of the federal government—the US Agency for International Development (USAID)—both saw the program as a way to demonstrate American superiority, win allies, and limit communism's spread. DDT was crucial to this objective. Its quick-fix approach to the problem of malaria promised an expedient way to secure allegiance to capitalism. This Cold War "proxy battle," as one observer called it, was hardly a secret. A political cartoon in an Italian newsmagazine showed a man whose spray gun washed the Russian hammer and sickle off the garments of a woman dressed in rags.

Russell, for his part, was caught up in his own country's anti-communism and sense of superiority. When he heard a report that the Soviet Union had already "liquidated" malaria, he scoffed. When a book on malaria, published in Russian, arrived at the program's Geneva offices, Russell noted that it referenced only scientists behind the Iron Curtain. It was as if no other scientist in the world (himself included) had ever studied the disease, he thought. When he encountered anti-American sentiment on his travels through India—which was often—he knew that propaganda from Moscow was behind it. And he wasn't surprised that it worked. "The Indian people have always been surprisingly naïve," he wrote.

When he took the time to think it all through, however, Russell wasn't sure what conclusions to draw about the connection between malaria eradication and politics. On the one hand, he was convinced by the argument that people locked in poverty because of disease might see communism as the best means of securing food, land, and well-being. On the other hand, he worried that people in good health might actually "raise more political Cain than ever." He looked to Egypt and Israel, where malaria eradication had progressed quite far and where political unrest was as prevalent as ever.

But Russell was first and foremost a physician. What mattered to him most was that DDT prevented disease safely, and at little cost. In 1955 the entire program cost 23 cents per person per year to protect more than 230 million people the world over. Numbers like that, he believed, spoke for themselves.

•

As 1956 began, Russell was at home in Maine, writing papers and book chapters. The winter was harsh: fifty inches of snow fell in March alone. He traveled to Athens for a conference on the eradication program. Six countries were just getting started, twenty-three were in the early stages, fifteen were well advanced, and in nine, including the US, eradication was complete or nearly there. Although mosquito resistance to DDT was still a problem, using sequences of different insecticides seemed to slow its development.

The following year, 1957, brought another horrid winter to Maine. This time eighty-one inches of snow fell in January alone. Russell's diary entries grew perfunctory and thin as the eradication program's challenges began to mount. Seventy countries were now participating in the program, and eradication within a decade seemed possible in the Americas, Europe, and parts of Asia and the Middle East. "But it will be a long time coming in Central Africa, and parts of Asia and Oceania," Russell wrote.

The problems varied. In the British Cameroons 20 percent of houses reportedly couldn't be sprayed because no one was home. In Western Sokoto the target anopheles species was biting people outdoors, not indoors. In several countries the malaria parasites were showing resistance to several different malaria drugs. All of these factors were combining to drive up per-capita costs. "Central

Africa," Russell wrote, was proving to "be the most difficult problem to solve."

In the end, the program would be terminated before campaigns in sub-Saharan Africa, next on the list, ever got fully underway.

•

The malaria-eradication program would continue through the 1960s, spraying hundreds of millions of people's homes with DDT before it became clear that the program's end goal simply wasn't feasible. More and more mosquitoes would become resistant to more and more pesticides. More and more parasites would become resistant to malaria drugs. The anophelines' behavior would vary so much from place to place that in some areas, it wouldn't even matter that DDT was still effective: either the mosquitoes never landed on the sprayed surfaces, or they landed too briefly for it to matter, or the surfaces were so complicated, like thickly thatched ceilings, that it was impossible to fully coat them with DDT. In some places, teams would spray and spray, yet malaria cases would just tick up.

But at the end of the 1950s, the program was still counting up its successes. By the end of 1958, it had reached more than half the global population at risk of malaria. Cases continued to tumble in Indonesia, India, Ceylon, and elsewhere. In Ceylon the number of malaria cases for the entire country would fall to just seventeen in 1963. The program's advocates considered its achievements spectacular.

But the program also had its critics, who would become more vocal over time. They saw the program as another form of colonialism, a militaristic campaign run by global and local elites. They criticized the WHO for inaugurating a purportedly global program before most of the world's poorer nations had even become members. They chastised the WHO for giving more weight to proposals from wealthy countries than poorer ones, and for putting member nations in a position where they effectively had little choice but to join the program. They saw Russell's tour as an act of deception and coercion: he promised poor nations that if they spent more money on malaria, they would get more foreign aid in the short term and be able to spend less of their national budgets on health care in the long term. Neither promise turned out to be true.

Russell likely would have argued differently, but at the end of the 1950s he wasn't asked to. He did stay on top of the various criticisms of DDT he was hearing in the United States—that it caused such health problems as polio, Virus X, and even cancer. He read Biskind's papers, he looked into the claims as thoroughly as he possibly could, and then he didn't pay them any further mind. The evidence of any harm from DDT seemed slim. Hundreds of millions of people all over the world slept at night in homes sprayed with DDT, and in his view they were undeniably and incontrovertibly healthier for it.

PART II

Chapter 11

DON'T CALL IT A POISON

When the phone rang early on June 5, Marjorie "Hiddy" Spock had been hoping for better news. Instead, Mr. Kilgallen from the local agriculture office told her that the spray planes going out the next day would pass near her home in the morning. Despite her many requests, he just couldn't promise that her two-acre property in Brookville, Long Island, would be spared. Spock hung up and acted fast. She packed her partner, Polly, into the car and drove her to a hotel in New York City. Back home that afternoon, Spock pulled out the massive plastic sheeting she had bought just in case. She spread it over their vegetable and herb gardens, the fruit trees in their orchard, and the dozens of berry plantings all over the property. It took ten hours, but by midnight everything was covered.

The next day, at work, Spock began to worry about her plants wilting under the plastic's heat. At lunch she rushed home to pull off the sheets. When she got there, they were dry. She felt a wash of relief; they hadn't been sprayed after all. Two days later, though, she was having breakfast when she heard the roar of an airplane engine. Stepping into the yard, she saw a low plane passing out of view. Brookville's tall, woodland trees meant she couldn't track it very far. But the plane returned a few minutes later, again a few minutes after that, and then again and again. She counted. The plane passed overhead fourteen times before it disappeared, leaving the smell of kerosene thick in the air.

She knew that smell meant DDT.

The plane was one of dozens commissioned to spray Long Island with DDT in the spring of 1957. The pesticide was the cornerstone of a federal program targeting the gypsy moth caterpillar. The hungry, fuzzy-headed caterpillars had been imported to the US back in the nineteenth century by a hobby scientist hoping to start a silk industry in Massachusetts. The industry didn't take off, but the caterpillars did. They multiplied at such a clip that by the 1880s, local residents said they could hear the sounds of them crunching and the patter of their excrement hitting the ground on summer nights. The caterpillars spread from there, feasting on the leaves of shade and fruit trees. In particularly bad summers, they stripped whole swaths of New England forest and woodland bare.

For decades, entomologists tried to limit the caterpillars' damage and spread. They lit them on fire, hand-picked their eggs off trees, imported predator insects from Europe, and hit infested trees with lead arsenate. The caterpillars persisted. Then, in the late 1940s, a few scientists in Pennsylvania tried spraying affected trees with DDT. The caterpillars disappeared. Successful experiments in Massachusetts and Michigan followed. Emboldened by the results, in 1956 the USDA announced a program that would spray three million acres of northeastern moth habitat over five years. Spray season would begin each May, by plane, to reach eggs deposited high in woodland canopies right when they were hatching. This, the USDA promised, would eradicate the caterpillar once and for all.

When Spock heard about the spray plans, she called the USDA's local plant and pest control division in nearby Hicksville right away. The office's supervisor, Lloyd Butler, was out, so she left the first of what turned out to be many messages with his secretary. She spent the next six weeks writing letters and calling USDA offices on Long Island and at the state capitol in Albany. She begged not to be sprayed. She offered to prove that her property had no caterpillars. She asked for mercy: Polly was sick, and their spray-free vegetables, fruit, and dairy products were vital to her health. But no one—except maybe Kilgallen—seemed to care. As the spray date approached, she wasn't quite sure what to expect, but she was not at all prepared for fourteen unannounced passes of a spray plane in a single day.

Not long after that June morning, the damage Spock feared began to set in. Their four hundred feet of pea plantings browned and died before bearing fruit. Raspberry and strawberry plants withered. Beet, chard, and lettuce leaves were shot through with holes. Pests that Spock had never seen before in their garden—potato scale, red spider mites, and slugs—suddenly turned up in droves. The wood thrushes, once numerous around their home, were gone. She despaired over what Polly should eat and how to keep Polly's illness from coming back. She sent a sample of their soil to a laboratory she knew of in the city to have it tested for DDT. She later sent some beets, too. The spray, she thought, was a "poisonous trespass." Friends told her not to, but she had made up her mind: she was going to sue the government.

•

Spock had found herself on Long Island with Polly after two decades of teaching and running some of New York City's finest private schools. She had grown up as the tall, hardy, and spirited second-youngest child of a large and well-to-do New York family. Her father was the esteemed attorney for the New York, New Haven, and Hartford Railroad. Her older brother Benjamin had grown up to become the nation's most famous pediatrician. Spock herself was admitted to Smith College after high school, but she had gone to Dornach, Switzerland, instead, against her family's wishes, to study at an institute run by philosopher Rudolf Steiner. She fell in love with Steiner's teachings about spirituality, humanity's connection to the Earth, and human development (Steiner created Waldorf education), and she stayed for years. She had run into Polly there—they had known each other as teenagers in New York—but it was just a passing encounter. When Spock's father died in 1931, she left Switzerland and went back home to New York.

Spock ran into Polly again two decades later, at a Waldorf conference in New Hampshire. Spock was astounded by how little her old friend resembled her younger self. Polly, in her early forties then like Spock, was pale and shockingly thin. Spock was sure she could not have weighed a hundred pounds. Polly shared the story of her illness: for years, she had suffered gastrointestinal symptoms so severe they confined her to bed for days at a time. Grocery and restaurant food seemed to trigger it. Unable to eat, she had grown thinner and weaker. But Spock still found her charming, with a radiant warmth. Drawn to her, she offered to get Polly a teaching job and help her build up her strength.

They bought the house in Brookville together in 1951. On a physician's recommendation, they began to grow their own food. Polly spent less and less time in bed. She put on weight. Her complexion warmed. Spock believed she was healing. She could not imagine losing those gains. When she decided to sue the government, it was to protect Polly. But it was also a decision they made together, because it was Polly's family money they would use to cover the costs.

It wasn't terribly difficult for the two women to find a lawyer. But it surprised Spock how easy it was to find other Long Islanders eager to sue. The retired curator of ornithology at the American Museum of Natural History in Manhattan, Robert Cushman Murphy, had a home not far from theirs, in Oyster Bay. He was incensed that the spray had killed the minnows he had bred to eat mosquito larvae in his ponds. He joined Spock's suit and knew of others who would, too. He connected Spock with banker J. P. Morgan's daughter, Jane Nichols, who owned dozens of acres of cropland and dairy pasture in nearby Cold Spring Harbor. He also put Spock in touch with Teddy Roosevelt's son, Archibald Roosevelt, an army colonel and wildlife defender who lived in the same moneyed enclave. Before long, Spock had more than a dozen enraged plaintiffs and lawyers from a firm of Princeton grads in the city.

Spock took the lead organizing the plaintiffs into the Committee Against Mass Poisoning, CAMP for short, headquartered in the basement of her and Polly's home. When school let out for the summer, she researched DDT full-time, collecting scientific papers, government reports, and newspaper and magazine articles about the pesticide. Polly's illness began to make more sense to her than ever. One doctor she found, Morton Biskind, had documented similar symptoms in hundreds of his own patients. She wrote to him right away to ask if he would be willing to explain DDT's hazards in court. He was happy to share his papers, he said. But he wasn't interested in being a witness. He had testified in the Delaney hearings several years before, he told Spock, and the attacks on his reputation and credibility that followed were not something he cared to face again.

Biskind pointed Spock to some of his colleagues instead. Gradually, she amassed a list of impressive doctors and scientists willing to testify. Two were physicians convinced, like Biskind, that pesticides were making their patients ill. Another had actually measured levels of chlorinated hydrocarbons in his patients' bodies since 1950 and had found them steadily climbing. But Spock was most excited about two other witnesses: Malcolm Hargraves, a physician at the Mayo Clinic whose research showed that pesticide exposure appeared to cause leukemia, aplastic anemia, and Hodgkin's disease; and Wilhelm Hueper, a National Institute of Health scientist who was one of the nation's top experts on chemicals harmful to human health. "With the scientific data now assembled," Spock told a friend, "we should win in a walk."

Later that summer, Spock's lawyer, Roger Hinds, introduced a motion in the Eastern District Court of New York seeking a permanent injunction against DDT aerial spraying. It named Lloyd Butler, New York State agricultural commissioner Daniel Carey, and US secretary of agriculture Ezra Benson as defendants. Lawyers for the defense asked the judge to dismiss the motion, arguing that the spraying was already complete. Hinds argued for a trial: Long Island's spraying might be complete, he pointed out, but it was just the start of a five-year program, one that violated the Fifth Amendment by depriving plaintiffs of property and possibly lives without due process of law and by taking their private property for public use without just compensation. It also violated the Fourteenth Amendment, he argued, via illegal trespass on private property. The judge agreed to hear the case. And in February 1958, Spock and her plaintiffs went to court.

•

A blizzard hit the Northeast the night before the trial, blanketing Brookville under several feet of snow. When Spock woke that morning, the streets were impassable. Their house was more than a mile from the closest train station, but she didn't dare miss their day in court. She had arranged to take two weeks off of work, and Hinds had encouraged her to bring to court the masses of material evidence she had collected: scientific papers, government reports, news items, and letters from citizens opposed to the spray campaign in other states. She tucked a thick stack of files under each arm and trudged to the station through the still-falling snow.

When she arrived at the US District Court in Brooklyn, just minutes before Judge Walter Bruchhausen called for order, Spock was nerve-racked—but also optimistic. Three judges who had heard the plaintiffs' motions to date had all ruled in their favor, allowing the case to come this far. The case had been given calendar preference too, putting it ahead of a two-year backlog of cases. And a young reporter, William Longgood, had already written about their case for the *New York World-Telegram and Sun*, making her hopeful that other reporters would show up that day, too.

The morning began with Hinds's opening statement. Spock listened to his familiar argument. He offered the court an analogy, invoking the national debate over prohibition. Never, he said, would anyone have suggested that alcohol consumption be compulsory. But the plaintiffs had been forced to consume DDT "against their will." The government had also abused its authority by trespassing on private land and had caused "irreparable injury" in doing so. DDT surely had its advantages, he added, but they hadn't been duly weighed against its harms.

The defense, in response, pointed to a 1928 Supreme Court ruling that upheld the Cedar Rust Act of Virginia, affirming state authority to cut down trees on private property to prevent the plant disease's spread. "We are not going further than that," said Assistant US Attorney Lloyd H. Baker, representing the Department of Agriculture. Like cedar rust, the gypsy moth caterpillar was capable of causing "serious economic loss" on the order of "millions and millions of dollars," Baker said. DDT spraying had certainly killed birds, fish, and aquatic insects, he granted, but their populations had returned to normal within a few months. And the spraying may have forced people to consume DDT, but DDT was already widespread in the American diet. The average person in the US already had DDT in his or her fat, and government scientists had shown that this was harmless.

DDT was not the problem, Baker argued; the problem was "reactionaries" fighting every new scientific advance from anesthetics to vaccines to X-rays. "But science," he told the court, "will not be stopped."

Spock soon noticed, with disappointment, that no other reporters had shown up. Given the snow, it wasn't terribly surprising. She spent the day taking her own detailed notes, a decision that would later nudge the course of history her way. But that day, all she thought of was CAMP, which had promised news of the trial to interested parties across the Northeast: home owners who had spoken out against gypsy moth spraying in Massachusetts and Vermont, her friends at *Organic Gardening and Farming* magazine in Pennsylvania, readers who had donated to their cause, and her farming mentor and fellow Steiner acolyte, Ehrenfried Pfeiffer, who ran a biodynamic farm in Chester, New York. Spock had a typewriter and a brand-new Thermo-Fax machine in her basement, and as she made her way home from court that evening, she made plans to head downstairs after dinner to type up the first edition of what she planned to call "Today in Court."

Polly was feeling ill again when Spock got home, but she stopped Spock with some news. A literary agent named Marie Rodell had called the house. She represented the nature writer Rachel Carson. Spock knew and loved Carson's work: her 1951 book, *The Sea Around Us*, and her more recent book, *The Edge of the Sea*, with its passages that captured with vivid brilliance the craggy coast of Maine, where Spock had spent her summers as a child. It seemed unfathomable that Carson had called Spock's home. Rodell, Polly reported, had said that Carson was curious to learn more about their case and that she was especially interested in hearing about the specific evidence they had collected.

Spock's heart leapt. She pored through her materials, selected a handful of articles and reports, and copied them on the Thermo-Fax. Then she sat down to compose a letter to Carson. "We have marshaled some pretty solid scientific men and data, and are feeling confident," she wrote. But she wasn't so naive as to assume the case would be open and shut. Never one to waste an opportunity, she asked whether Carson, as such a respected authority on wildlife, would be willing to serve as a witness to strengthen their case. "I think you know how grim this struggle with the US government and the whole chemical industry is bound to be," she wrote. She sent off the packet and waited for a reply.

•

Carson had first tried to write about DDT back in 1945, before the war's end. She was a government aquatic biologist who aspired to write, and at the time she was working as an information specialist for the US Fish and Wildlife Service. Her job was to turn government research reports into press releases. After writing a release about DDT's effects on wildlife at the Patuxent Research Refuge in Maryland, she had tried to pitch an article about it to *Reader's Digest*. She thought it had all the elements of a broadly appealing nature story. Here was a wonder chemical with a potentially significant dark side; after all, you couldn't kill insects without harming the species dependent on them to survive. But the magazine wasn't interested. So she moved on to other things.

The idea of writing about DDT's effects on the balance of nature stuck with her, though, long after she stopped working for the government in 1952. In 1957 she learned of plans for two government spray programs, one targeting fire ants in the South, the other gypsy moths. Through her old government contacts, she talked to officials at the USDA and FDA who were concerned about the massive use of new insecticides planned for both programs. DDT would be used against the moths, and dieldrin, even more toxic than DDT, would be used against the ants. When she heard about the gypsy moth trial to take place on Long Island the following winter, she wrote to *New Yorker* writer E. B. White to see if he was interested in writing about it. Back in 1946, he had written a short article speculating that DDT might yield a future without fish, frogs, or birds. White said he wasn't interested, but he suggested that Carson propose an article to his editor, William Shawn. She did. Shawn was interested. So Carson asked her agent, Rodell, to secure whatever trial documents she could.

Carson pored through the materials from Spock, landing with interest on a forty-page summary of DDT's hazards in an obscure farming journal. The author, Ehrenfried Pfeiffer, was a Swiss-immigrant chemist turned farmer with 160 acres in Chester, New York. He ran a research lab in Spring Valley, New York; was a member of the American Chemical Society and the American Association for the Advancement of Science; and had experimented with DDT for three years. His article was a compendium of its problems, drawn from American and European scientific literature. It described fish kills following forest sprayings; DDT's accumulation in soil, milk, and butter; and the explosive growth of aphids, mites, and other pests in places where DDT spraying had killed the predators and parasites that normally kept them in check. The article referenced fifty-one papers by scientists at the US Fish and Wildlife Service. It reviewed dozens of studies on DDT poisoning and ingestion in people. And it concluded that assertions of DDT's harmlessness were doubtful. Pfeiffer knew well that "much medical and public health opinion" held DDT harmless, but it seemed to him, he wrote, "that this is not the case." After all, scientists now knew that a burn was not the only damage an X-ray or radiation could inflict.

The article and its six-page bibliography gave Carson a clear road map for exploring DDT's hazards. She made plans to track down the countless studies and reports it cited, and in the meantime, she wrote back to Spock. She regretted that she could not testify, she said, but she wanted Spock to know how grateful she was for everything she had shared. "Dear Mrs. Spock," she wrote, "you have been so enormously helpful to me, and apparently are so familiar with a vast amount of material." The "very excellent Pfeiffer paper," she added, was "a gold mine of information."

•

Over that first week in court, the plaintiffs called their witnesses, the defense cross-examined, and Spock took notes. One New York doctor's testimony about his sick patients was "magnificent," she wrote. When Hargraves appeared from the Mayo Clinic, "the defense blanched." Hargraves described a frightening rise in cases of aplastic anemia, lymphomas, and leukemias. He said he had diagnosed a "large group" of patients with insecticide poisoning caused by DDT, chlordane, lindane, and other new insecticides.

"Do many of these patients have psychosomatic disturbances?" asked Baker, following a now-familiar line of questioning.

"No," said Hargraves. He described their case histories, detailing their respiratory problems, weakness, swollen lymph nodes, and tumors.

Spock was delighted with how much valuable testimony the defense attorneys inadvertently extracted from him. "Mr. Hinds has suggested that we pay them for their invaluable contributions," she wrote.

Spock ended the first week of trial eager for the second to begin. The plaintiffs planned to present what Spock called their "real ace card": DDT's ubiquitous presence in milk. Their dairy farmer witnesses testified that DDT was still present in their cows' milk eight months after the spraying. Their physician witnesses testified that DDT had no place in the diet of infants and small children, that people sensitive to pesticides needed a source of milk free of DDT, and that DDT wasn't just in cows' milk; it was also in human milk. Spock could hardly wait for each day to unfold—this was exactly the scientific knowledge that deserved its day in court.

The defense, however, was prepared. DDT was found throughout the American diet, their witnesses testified. As a consequence, its presence in milk was neither exceptional nor worrisome. "It is no more harmful in milk than in anything else," Baker said.

And then he called Wayland J. Hayes to the stand. Hayes was a still-new CDC lab director when he had testified before the Delaney committee seven years before. Now he had nearly ten years at the CDC's Toxicology Section under his belt. He had published almost twenty papers on the toxicology of DDT and had written a chapter on DDT for a book edited by Geigy chemist Paul Mueller, who received the 1948 Nobel Prize in Physiology or Medicine for recognizing DDT's properties. He took the stand with confidence, poise, and a plan.

Hayes described experiments in which he fed DDT to male and female Sherman-strain rats for two years. Severe poisoning happened only when they ate 1,000 ppm. He described one study in which his scientists tested restaurant food for DDT and another study in which they measured DDT levels in the fat of four types of people: those who died before DDT was discovered, vegetarians, omnivores, and agricultural workers. People in the first group had no DDT in their fat; those in the last had the most. But "the critical thing," Hayes said, was what DDT stored in the body meant for people's health.

Hayes stood and went to a blackboard in the room. Picking up a piece of chalk, he drew several graphs to illustrate what he was about to demonstrate. It was widely known that DDT, like any material that stored in the body, accumulated over time, he explained. "What is not so well known," he said, "is that the storage of this material will eventually reach an equilibrium." He indicated a point in one of the curves he had drawn, where it began to plateau. "It will eventually reach the certain level and go no higher," he told the court. "This rather simple concept," he added, "is very important in interpreting any statements about the storage of DDT in man or any other organism."

He went on. In his study of men imprisoned in Atlanta and Tallahassee, which had been recently published in the highly regarded *Journal of the American Medical Association*, men ate up to two hundred times more DDT than the population generally consumed. Those men who ate the most DDT stored the highest levels of DDT, and its breakdown product DDE, in their fat. However, those levels appeared to reach a state of equilibrium in about a year; Hayes's graph showed a plateau. And at that point, he said, the men were all still in good health.

"I was strongly impressed with this witness," Judge Bruchhausen would later write.

As Hayes testified, Spock sensed the plaintiffs losing ground. Hinds attempted to regain it. He asked Hayes whether he had submitted the records of each prison volunteer as evidence. No, the files formed a stack six inches high, Hayes said, implying that such volume was too much to burden the lawyers with. Hinds noticed that while the plaintiffs' witnesses had testified that DDT storage in American bodies had increased since 1950, Hayes implied that it hadn't. But when Hinds looked closely at the reports Hayes had submitted as evidence, they didn't support the claim. Hinds requested that Hayes return for further questioning, but the defense declined. The defense then withdrew Hayes's reports and asked that only his oral testimony remain on record. Bruchhausen approved the request. In his decision months later, he would laud Hayes as the leading expert in the room.

Spock's optimism waned even further when their witnesses described living through the spraying and its consequences. Her fellow Long Islanders told of being drenched with spray on their walks to work, of finding dead birds on their property, and of finding dead fish in local ponds. One witness described the ruined swimming pools in her neighborhood, where even the children's playgrounds were "sprayed thick with oil and poison." Lawyers for the defense objected. They asked that the word "poison" be struck from the record. The judge sustained. Spock grasped what was happening. The word *poison*, once common, was no longer allowed in a court of law in which DDT was the defendant.

In the end, Longgood was the only reporter to cover the case. "The trial was long and bitter," he wrote. It seemed to him—and to Spock—that the fifty expert witnesses who had testified for the plaintiffs and the defense all agreed that DDT was a harmful chemical. Spock had started out with the hope that proving DDT's harm would be sufficient to win the case. By the time the trial ended in March, she knew it wasn't. The defense accepted that DDT was a hazard—one that people could and should tolerate. The judge seemed to agree. At best, Spock came to believe, her case would "stand or fall" on the government's right to spray private property.

•

Each day that Carson received a copy of Spock's "Today in Court," she thought she really had struck gold. She wrote back to Spock in gratitude. Spock replied with a never-ending stream of papers and reports for her to read, and she kept sending them after the trial ended. Sifting through it all, Carson learned about Biskind, Pottenger, Hargraves, and others who believed that DDT and other postwar pesticides were harming the health of the population. Their explanations drew on a vast published literature on the toxicity of the new pesticides. But no one seemed to have pieced it all together.

Meanwhile, DDT production was still climbing. US manufacturers were now producing nearly 100 million pounds a year. And DDT itself wasn't even the half of it. In 1956 US chemical manufacturers had produced more than 500 million pounds of synthetic organic pesticides, which included the organic hydrocarbons, such as DDT, and the more acutely toxic organophosphates, such as TEPP. Farmers sprayed them on growing monocultures of crops from cotton to tobacco to corn. Government workers sprayed them on vast acreages, less often to kill pests carrying human diseases and increasingly to kill a much broader definition of pest, such as gypsy moths and fire ants. The US was also applying pesticides abroad, too, its DDT the cornerstone of the World Health Organization's malaria-eradication campaign. Pesticides had never been produced or used on such a scale.

Working with Shawn at the *New Yorker*, Carson planned a 50,000-word article about the threat that poisons posed to people and wildlife. The scale of the problem was so massive, though, that even as she wrote, she knew the article was just a precursor to a book on the subject. Her friend Dorothy Freeman, with whom she spent summers in Maine, called it "the poison book." To close friends like Freeman, Carson admitted that she wasn't interested in treating the subject with any sort of journalistic fairness. When she shared this thinking with Shawn, he encouraged it.

"After all," he said, "there are some things one doesn't have to be objective and unbiased about."

Chapter 12

THE POISON BOOK

Marjorie Spock and her plaintiffs heard back from the court in June 1958, a year after her and Polly's property had been sprayed. The news wasn't good: Judge Walter Bruchhausen had ruled in favor of the defense.

Bruchhausen's opinion finally made it clear to Spock that her suspicions of bias were well-founded all along. The judge said he thought that the plaintiffs' expert witnesses were unable to prove damage to health. Because so many of them were "so strongly in favor of organic farming, without the use of chemicals," he said he believed that their testimony had to be "scrutinized." Meanwhile, the evidence had convinced him that the gypsy moth caterpillar was a demonstrable "evil" and that DDT was effective against it. For that reason, he found, the spray program was a "valid exercise" of the government's "police power"—that is, its authority to abridge private rights to serve the public good.

The ruling stung Spock. Bruchhausen's comment about organic farming particularly rankled her. It was "absurd," she said. "Most of the plaintiffs and many of the witnesses in scientific fields had never even heard of organic farming." She felt as if she and her plaintiffs had been brushed off like the very flies that DDT was meant to kill.

At their home in Brookville, she and Polly talked through their options. With Polly's money, they could afford to take things further, so they decided to appeal. While Roger Hinds prepared, Spock sent Rachel Carson more materials. "Dear Marjorie," Carson wrote to her that summer. "I feel guilty about the mass of your material I have here . . . you are my chief clipping service, and I do appreciate all that you do." When Carson asked Spock if she could also send along the full transcript of the trial proceedings, Spock didn't hesitate. She got a copy from Hinds and sent it off.

Sending Carson materials was the most productive and hopeful thing Spock thought she could do that fall, as she and Polly waited to hear whether their appeal would be heard. When Carson shared that she would be traveling up to New York to hear a talk by Malcolm Hargraves, Spock insisted they meet for lunch. She brought Carson a file full of clippings and a little handmade wren house for her five-year-old adopted son and grandnephew, Roger, whom Carson often mentioned. Spock, who worked with children but had none of her own, slipped a playful note inside for the boy. He loved it.

"They are using it!" Carson reported back to Spock that spring. "The wren babies are chattering in their little house." That same season, in April 1959, the court of appeals agreed to hear their case.

As Spock's relationship with Carson grew into a friendship, she and Polly made plans to visit Carson at her cottage in Maine that summer. Spock's mother had a house up there on the coast, a place she loved for its vital air and sparkling light. And as Spock knew from Carson's books, it was a place the writer also cherished. They met for lunch at the Newagen Inn, a colonial clapboard hotel on a Southport Island cape that jutted out into the sea. Afterward, Carson brought Spock and Polly to her home, perched on a cliff overlooking a wide sweep of bay. They walked down to the beach to explore the ocean life caught in its tide pools. Back at the cottage later that day, they pored through the materials Spock had sent.

Spock was inspired by Carson's extensive and painstaking research, her careful thinking, and her plans to condemn all of the widely used poisons. She decided the most useful thing she could do was entertain Roger so that her friend could have some quiet time to write. She took the little boy to an old military fort for an afternoon. Back in New York, she began sending Carson materials several times a week, interspersed with more gifts and notes for Roger.

In the fall, the appeals court upheld Bruchhausen's decision. Spock and Polly consulted again. They could afford to take things further yet. They told Hinds they wanted to appeal to the US Supreme Court.

•

Carson followed up on just about every lead in Spock's files. She wrote to Irma West, a physician documenting the deaths of farm-workers from Mexico for California's state health department. After reading Morton Biskind's testimony, Carson read his published work "avidly" and struck up a correspondence with him. After hearing Hargraves speak, she wrote to him, too. Biskind and Hargraves encouraged her to look closely at the work of physician Wilhelm Hueper at the NIH.

Hueper was a German-trained physician who had studied factory workers' health in Germany early in his career. While working as a pathologist in Pennsylvania in the thirties, he had visited Du-Pont's dye plants, where he warned management that the workers were using a chemical known in Germany to cause bladder cancer. DuPont immediately hired him to lead a company lab to keep track of such dangers. Two years later, they let him go. Hueper went on to write the first textbook on occupational cancer, a nine-hundred-page tome, and to found the Environmental Cancer Section at the Public Health Service's National Cancer Institute. He advanced a scientific approach that identified workplace carcinogens by responding to case reports brought forth by physicians and then evaluating them through experiments on animals in the lab.

The approach put him at odds with proponents of the new field of industrial epidemiology. In an effort to understand the nation's steep climb in cancer cases—especially lung cancer—in the 1930s, epidemiologists had increasingly used "case-control" and "prospective" studies to clarify the relationship between a workplace exposure and disease. Case-control studies compared people who had been exposed to something with people who hadn't. Prospective studies followed healthy people over time to determine which of their earlier exposures resulted in later disease. Case-control and then prospective studies had recently implicated cigarette smoking in the stark rise in lung cancer rates, but Hueper was circumspect. He worried that smoking studies foretold a future in which people's individual choices would be blamed for their cancer, letting industry off the hook. He worried that epidemiological results were too often taken to be universal truths, broadly applicable across an entire population. And he worried that industrial epidemiology left out too much; it assumed that carcinogens were confined to the factory, when in fact they were everywhere.

Hueper had said as much before the Delaney committee in 1952. Years later, Carson read his testimony. Although there was no evidence that DDT caused cancer in humans, he stressed that it might take ten to fifteen years to see such a connection and that that research simply hadn't been done. Decades later, he'd be proven right.

In the meantime, Hueper's ideas made Carson think she needed to look more closely at Wayland Hayes's work asserting DDT's safety. But when she looked at the notes she had taken when she had borrowed the trial transcript from Spock the summer before, she realized she had taken down very little of what Hayes presented in court. "I wonder whether you or Mr. Hinds could now spare that particular volume of the testimony for a short time?" she wrote to Spock in the fall of 1959. "I should like to make the most of the fact that the detailed data in his often-cited experiment were not made part of the record and that Dr. Hayes himself was not available for cross examination."

Hayes had conducted a prospective case-control study on DDT consumption, dividing his imprisoned volunteers into those who ate DDT and those who ate something that only looked like it. But when rereading his testimony and his published work, Carson became convinced that he had obfuscated evidence. To start, he had exaggerated the number of men in the study. In court he had said it was fifty-one, but his published paper showed that this was the number who participated on the first day. Only a "handful," she found, had actually stayed in the study until the end. Just twenty-six men consumed DDT for a year, and only some of these men kept going for the full eighteen months. Some of the men who left the study no longer wanted to participate, some left prison, and some refused parts of the study, like the biopsy required to measure their fat storage. Moreover, Hayes had claimed in court that the men reached equilibrium in a year, but in his own publication he had had to acknowledge that further research was necessary to support the conclusion. "One of my delights in the book," Carson told a friend, "will be to take apart Dr. Wayland Hayes' much cited feeding experiment."

Yet the more she read, the less this particular point seemed to matter, as something else began to occur to her. Hayes's other published studies proved that people stored DDT. Hueper, meanwhile, noted that potential human carcinogens were everywhere. Carson thought back to a host of published studies Biskind had pointed her to, which now seemed newly relevant. Animal studies showed that DDT interfered with the synthesis of proteins. Proteins were the building blocks for enzymes, catalysts for the biochemical reactions governing everything from breathing and digesting to moving and thinking. Research on nerve gases, to which some of the postwar pesticides were closely related, showed they affected the specific enzyme cholinesterase, an enzyme required for nervous-system function. The deeper Carson dug, the more studies she found on enzyme disruption. By late 1959, she had a long list of pesticides that disturbed the function of the enzyme cytochrome oxidase, in particular, which was involved in cellular respiration, the process by which cells turn oxygen into energy. Ehrenfried Pfeiffer's paper, she recalled, had also noted the connection. She wrote to Biskind, who pointed her to a published paper describing an experiment in which researchers had caused cancerous changes in tissue samples by restricting and restoring their oxygen supply.

The pieces seemed to fit together like a puzzle, one no one else had bothered to figure out. "A great light is breaking in my mind," she told Biskind. "Perhaps I shall be a fool rushing in where angels would not tread, but I think this possible mechanism should be suggested as one means by which the cell may be converted to the cancerous way of life." DDT's greatest harm, she feared, might well be an ability to cause cancer that wouldn't show up for years or possibly decades—as Pfeiffer had implied, its harm might simply be concealed deep inside living things. Biskind replied with yet more references to track down. "My respect for Dr. Biskind deepens," she wrote to Spock. "For all that I now consider most important I was led to by him."

But as the pieces came together, Carson struggled to find time to write. She was caring for her elderly and ailing mother as well as little Roger, with his constant need for play and attention. Early in 1960, she was diagnosed with an ulcer, and then she contracted a serious case of viral pneumonia and a sinus infection. On top of that, fewer of the government scientists once willing to speak with her were returning her calls. Carson was increasingly confident that she had arrived at a hidden truth. "I suspect that the great hazard is not so much toxicity as genetic and carcinogenic effects," she wrote to Spock. Toxicity accounted only for short-term damage, what happened to organs or systems when an organism was poisoned. But what if the damage was subtle, like an alteration of genetic material or a triggering of cancerous growth that might not be noticed for years or for even a generation? She was doing all she could, she told Spock, to write at a "furious pace."

•

The prospect of appealing to the Supreme Court was little comfort to Spock, who was despondent about her and Polly's homestead in Brookville. The DDT, she knew, would remain in their soil for years. She and Polly packed up the house and purchased a 140-acre farm in Chester, New York, adjacent to Pfeiffer's. Spock gave notice at the Waldorf school where she worked. She was fifty-five years old and just starting her life as a full-time farmer.

But the spring of 1960 brought bad news in one form after another. She received a letter from Carson detailing a litany of health troubles, and her heart absolutely sank. "I am almost tongue-tied with shock upon reading your letter recounting the list of your present health afflictions," she wrote to Carson. "Our hearts are deeply stirred, and we wish our *hands* could do something to help." She made plans to send healthy food from their farm and Pfeiffer's, along with more gifts to lift Roger's spirits and keep him occupied.

Then William Longgood published a book, *The Poisons in Your Food*, based on his reporting on their trial. Reviewers tore it to shreds. Esteemed nutrition scientist William Darby labeled it "muckraking" of the worst sort and lambasted Longgood for quoting "cult leaders" over "true scientists." Spock asked Carson if she had seen the reviews. She had. She guessed that the New York State Department of Agriculture had it out for Longgood. What neither she nor Spock knew was that the directors of the Manufacturing Chemists' Association had quietly launched a targeted attack on him, carried out by public relations firm Glick and Lorwin, specialists in "health" communications who mailed copies of Darby's review and "fact sheets" to a long list of opinion makers.

The next piece of bad news came from Hinds: the Supreme Court had declined to hear their case. Small comfort came in the form of a strongly worded dissent from Justice William O. Douglas. "The issues involved in this case are of such great public importance that I record my dissent," Douglas wrote. He acknowledged the "sharp dispute" among experts but pointed to the "alarms" raised by "responsible" parties. DDT's possible connection to leukemia and other diseases concerned him, as did its effects on birds, including the bald eagle, recently reported by the *New York Times*. He cited Biskind's work and Longgood's. "I do believe that the questions tendered," he wrote, "are extremely significant and justify review by this Court."

But Douglas's dissent made no difference. Spock and Polly's legal fight was over. They had lost. Spock shared the news with Carson. Though sick, her friend was still writing. For Spock, that meant the war against spraying was far from over. "Rachel's pen was by all odds the strongest means of stopping it," she believed. She resolved to do whatever she could to continue to help.

•

Carson didn't tell Spock right away, but her health had in fact only worsened. She found several growths in her left breast that spring and was forced to delay surgery when she contracted another viral infection. When she finally went in, the doctors removed two tumors and performed a radical mastectomy, removing the lymph nodes and much of her pectoral muscle on her left side.

A week after she left the hospital, in the spring of 1960, Carson wrote to Spock. She wanted only "special friends" to know of her illness, she wrote, lest the news appear in the literary gossip columns. "Too much comfort to the chemical companies," she grimly joked.

Over the next twelve months of attempted healing, Carson continued to work as her body fought her. The following spring, a staph infection in her knees and ankles confined her to her bed. She hired an assistant and dictated. When she recovered, she learned that the doctors who had operated on her breast had not told her the truth about the tumors they had removed: they were malignant, not benign. She shared the news with Spock, who sent hamburger meat and butter from a friend's farm. She asked Spock how she could get more—for Roger, who had loved every bite.

Still, she worked. After speaking with Hueper and reading one of his more recent papers, Carson had landed on yet another potential cancer mechanism. Hueper had argued that a chemical could be an indirect carcinogen by causing changes in an organ or tissue that made it possible for some other factor, such as a hormone, to produce cancer in the body. She ran the theory past Biskind, but he wasn't entirely familiar with it. She pointed him to a paper Hueper had published on hepatomas, cancerous liver growths, in rainbow trout. Curious, Biskind tracked it down and read it, and then quickly wrote Carson back. "The authors unequivocally include DDT among the carcinogens!" he wrote, thinking only of what this meant under the Delaney Clause. "Since this comes from the National Cancer Institute, the FDA is required by law to establish a tolerance of zero for DDT in all foods."

Carson had come to think of her still-developing chapter on cancer as one of the most important parts of her book. It described the nation's shocking climb in cancer cases. It described Hargraves's findings on rising rates of leukemia. It described how many pesticides, including DDT, caused cancer in laboratory animals. It offered two theories on how pesticides might cause cancer. In one, the destruction of cellular respiration stripped cells of their vital source of energy, turning them potentially malignant. In the other, damage to an organ, such as the liver, upended the body's balance of hormones, such as estrogens, which then reached levels capable of triggering cancer in yet other organs. With so many cancer-producing chemicals in foods, water supplies, and the environment, she wrote, it was also possible that some, like the ubiquitous hydrocarbons, were capable of acting together in the body to cause cancer. Given that, she asked, "What then can be a safe dose of DDT?"

By the time Carson was done drafting the book, Hayes warranted only a passing mention as one of the scientists who contended that DDT's storage in the body reached equilibrium. She swept past that point to note that everyone had DDT and its metabolites stored in their fat. The amount ranged from 5 to 7.4 ppm in most people, she wrote, and in "workers in insecticide plants as high as 648 parts per million!" The main question, to her, was not whether stored DDT plateaued but what the stored chemical was capable of doing to the body: on its own, in concert with so many other chemicals, and over the long term. A related question was how to define the long term. One of the scientists she interviewed raised the possibility that carcinogens might in fact act on fetal tissue in pregnant women. She quoted him in her cancer chapter. "We may be initiating cancer in the children of today," he said. "We will not know, perhaps for a generation or two, what the effects will be."

The *New Yorker* ran Carson's articles in June 1962. Her publisher, Houghton Mifflin, released her book in September. After much deliberation, it bore the title *Silent Spring*, two words that would ultimately become a cultural touchstone for the sea change that followed the book's release.

But in the meantime the book was attacked, far more harshly than Longgood's was. "Silence, Miss Carson," read the title of Darby's review. Darby said Carson had exaggerated evidence, quoted scientists out of context, and deliberately blurred the distinction between the hazards of occupational exposures and residues in food. Agriculture Secretary Benson called her a spinster and a communist. The science editor at *Time* was relatively gentle when he said the book "does not approach the literary level of *The Sea Around Us*" but that its fans were not *all* "faddists and hysterical women."

•

Spock was certain that the attacks on Carson came from "the pesticide interests and their hangers-on." She didn't have the proof herself, but in fact the MCA directors had been planning their defense, even joining forces with related industries, since Longgood had published his book. Spock and Polly read the "ferocious and scurrilous" reviews, infuriated on their friend's behalf. Spock could only imagine how hard it all was for Carson.

She and Polly watched as Carson made a handful of public appearances: in a televised broadcast on CBS; before Congress, twice, as the Senate held hearings in response to her book; and in a forum on pesticides in San Francisco, to which Carson traveled in a wheelchair, so weakened from the disease she still kept to herself. Spock could see that her friend was pushing relentlessly and desperately through illness and exhaustion.

Spock got a call from Carson late the following year, when things had quieted a bit. Carson told her she was working on a new book, a children's book about nature, inspired by her time with Roger. But she never finished it. A mutual friend called Spock the following spring to report that Carson was gravely ill.

It was April, and Spock and Polly were planning a drive up to Maine, where they had bought a new farm on the coast, near Acadia National Park. They drove past the turnoff to Carson's summer home, recalling the time they met her for lunch at the Newagen Inn and watched her come around the bend, unhurried and slight.

"When we knew Rachel better, we realized how typical it was of her to keep her own way in everything," Spock recalled. The next afternoon, she and Polly heard on the radio that Carson had died.

Spock sat down to write what she called a "portrait" of her friend. "There was something in her nature that made her immune to the paralysis of will that afflicts so many in our time," she wrote. "To see what needed to be done was, for her, to find the strength to undertake it." She sent the draft off to Rodell, Carson's agent, who filed it away.

Not long after, Spock sat down at her typewriter again, this time to write to the local papers. She and Polly had moved to Maine to get away from government sprays. But the state's forest commissioner had just announced a plan to spray thousands of woodland acres with DDT. The target of the proposed spraying was yet another forest pest: the spruce budworm. Spock composed a letter, determined to stop it.

Chapter 13

POISONED IN THE FIELDS

Sixteen-year-old Jessica Govea had a voice that captivated everyone who heard her sing. The gift surprised no one: music ran in the family. Her father, Juan, had studied at the Conservatorio Nacional de Música in Mexico City as a teenager. In his early twenties, though, he had traveled up to the border to apply for work on the other side, through the new guest-worker program with the United States. Young, strong, healthy, and unmarried, he made the cut.

Doctors had examined him up and down, tested him for tuberculosis and venereal disease, vaccinated him, and then covered him from head to toe in a white powder to kill lice: DDT. The men who deloused him and his fellow braceros, or "arm men," wore uniforms and masks to protect them from breathing in the dust. The braceros stood naked. "They fumigated us," said one, "as if we were a herd of cattle, goats, hogs, dogs."

Govea's father was sent to Arizona, where he worked on a track gang for the railroads. The work eventually took him to Hanford, California, in the rural Central Valley. The days were so long, the wages so chiseled, and the room and board so insupportable that the braceros in his gang marched up to the Mexican consulate in Fresno, more than thirty miles north by foot, to ask for help. The consul told them to solve the problem for themselves.

The encounter stayed with Juan when he met Margarita de la Rosa, whose family lived in tents on the orchards where they picked.

Juan and Margarita married and settled in a Bakersfield barrio named Little Okie, after the Dust Bowl migrants who had lived there in the thirties. And the encounter was with him still when he and Margarita met a former farmworker, Cesar Chavez, at a community meeting.

Chavez worked for a group called the Community Services Organization. The CSO was a project of the Industrial Alliances Foundation, founded by a community activist and organizer named Saul Alinsky, worlds away in Chicago. Chavez spoke in a quiet, steady voice of the things that the CSO had done in other communities: fighting police brutality, bringing basic services into the barrios, helping people become citizens or apply for disability and pension programs, and organizing farmworkers to protest when new braceros were brought in at wages so low they displaced locals from their jobs. Eager to bring change to Bakersfield, in 1955 Juan and Margarita helped Chavez found a local chapter of the CSO.

Their daughter Jessica was eight when she first heard Chavez talk to a group of farmworkers in the backyard of her family's bungalow. Everyone came to the meetings: mothers, fathers, and children. As the adults spoke, she listened. They discussed the things they wanted to change: spoiled food, crowded and filthy housing, long hours, and low pay. Also, dangerous conditions: unrelenting sun, rickety trucks, reckless crop dusters, no bathrooms or running water, and the poisons sprayed on neighboring fields and the ones in which they worked, picking everything from artichokes to walnuts.

In the summer of 1963, poisons were on many workers' minds, and not because of *Silent Spring*. Thousands of farmworkers, mostly men, had migrated up the Central Valley to Stanislaus County for piecework in the peach orchards near the town of Hughson. Peaches had to be picked by hand over a six- to eight-week period, the men working fast in the baking heat. They moved tall ladders from tree to tree, reaching in through branches and leaves to get every last fruit in the area's 240,000 acres of orchards. But that peach season, nearly a hundred men fell ill at once with vomiting and diarrhea, debilitating weakness, blurred vision, and pounding headaches. A number were sent to the hospital. One died.

Local doctors attending to the men notified the California Bureau of Occupational Health. The bureau sent a team of health officers to investigate. Peach pickers were suffering "a rapidly developing and serious epidemic of poisoning," they concluded. The culprit was parathion, sprayed a week to a month earlier, following all recommended guidelines. The men had absorbed it gradually through their skin. As they reached through the branches day after day, week after week, the residues on the leaves and fruit had entered their bodies bit by bit, until suddenly they had enough of the chemical inside them to trigger the symptoms of acute poisoning. All of the published scientific literature claimed that parathion disappeared rapidly after spraying, the officers noted, and that it therefore posed no hazard to people working in the fields well after it had been sprayed. But the epidemic in Hughson disproved the idea. Growers and foremen often wrote off similar epidemics as heatstroke or food related, but the bureau was now certain the chemical had been poisoning California farmworkers for years.

Govea and her family never saw the health officers' report, released to the governor and later published in a top medical journal. But they didn't need to. Govea had started working in the fields with her parents when she was four, hauling cotton. She carried a twenty-five-pound bag her mother made for her because the hundred-pound bags were too big. She had grown up picking cotton, grapes, onions, plums, and potatoes alongside her parents and little brothers. She had only recently stopped working in the fields to focus on school and the debate team because her parents insisted that she prepare for college. But she would never forget the sight of the crop dusters flying low over fields adjacent to those where they picked, the smell on the grapes, and the feeling on her skin at the end of the day. It "would itch and burn," she said. And she was sure it wasn't because of the sun.

•

Health officer Irma West's entire career, she sometimes thought, was characterized by pushing through resistance, from getting into medical school at a time when few admitted women to getting her job at the health department as a divorced mother of one. The hiring team at the department had wondered why, as an attractive woman still in her thirties, West wanted to work instead of getting remarried and having more children. After accepting the job, West made a point of working hard and dressing "stylishly," she told friends, so that no one would take her for the "typical woman doctor," unattractive and unmarriageable.

With an undergraduate degree in chemistry and a medical degree, West joined the department's Bureau of Occupational Health, assigned to review doctors' reports coming in from across California to determine which diseases and deaths were "occupational" in nature. She had quickly noticed a growing number of poisonings in the state's booming agricultural sector, so she wrote up a notice for physicians who worked in farming areas. Far away from cities and factories, they might easily conclude that they had no role to play in preventing occupational disease, she wrote. But with the changes in agriculture, especially the rapidly increasing use of what she called "the most hazardous materials yet used as pesticides," they needed to know how to recognize and treat resulting poisonings now happening in the field. "It was a hard sell," West said.

The pesticides she first warned of were parathion and TEPP, organophosphate pesticides related to the nerve gases developed during World War II. Organophosphates worked by blocking or destroying the enzyme acetylcholinesterase, which helped signals move from one nerve to another; when organophosphates knocked acetylcholinesterase amounts off kilter, the nervous system went into a potentially deadly state of overstimulation. That state could be reversed with injections of the drug atropine, but it was notoriously toxic, causing a long list of side effects from rapid heart rate to loss of balance. West acknowledged doctors' resistance to the drug and then pressed them to use it anyway.

West had also been curious to know just how extensive the pesticide poisoning problem was. California's agricultural sector was enormous, she knew. Whereas other states produced just over a dozen agricultural products at best, California produced more than two hundred, from tomatoes to cotton. It grew more than 40 percent of the nation's vegetables. To do so, the state used more than 20 percent of the nation's pesticides and most of the nation's braceros. In the late fifties, West's office had gotten word of a series of unexplained deaths among the workers. The men woke in the middle of the night, after a long day of toil in the field, thrashing from what their bunkmates took to be violent nightmares. Then they died. Local doctors called them "dream deaths" and surmised that narcotics, heart attacks, bad food, or "voodoo" might be to blame. West had wondered if the real cause was insect-killing chemicals.

In 1957 West attended an annual meeting of state coroners so she could describe the deaths and ask for help identifying others. She also asked the gathered coroners to investigate such cases for what she vaguely called "toxic elements"; she didn't want to clue them in to her hypothesis. Two days after the conference, she had twenty reports from coroners across the state. Many of the deaths remained unsolved mysteries, coroners' reports turning up no detectable poisons. But West knew that meant nothing. She began collecting reports of known pesticide poisonings from doctors and coroners. She studied them alongside reports from hospital emergency rooms and work injury reports collected by the California Department of Industrial Relations. She kept count. She was promoted to bureau chief. Still, she kept counting.

Finally, five years after her first warning to physicians, she had a crystal-clear picture of the extent of the problem. Pesticides were clearly poisoning a thousand people in California each year. Roughly ten of them died, mostly farmworkers and children. And poisoning epidemics like the one in Hughson were becoming more common. Her department put out a press release, and a few newspapers ran small news items in response. But she worried that the problem wasn't getting the attention it deserved. And as she talked to farmers, growers, and pesticide salesmen, it became abundantly clear why none of them wanted to do anything differently. They all said they didn't want to unnecessarily alarm farmworkers, but she could see the real issue: it cost less to risk worker illness than to prevent it.

And then, in 1962, there were her words in Rachel Carson's new book, *Silent Spring*, in a chapter titled "Elixirs of Death." Carson didn't name her; she referred to West only as a "medical authority." But it didn't matter. As someone who had been issuing warnings about pesticide dangers for years, West suddenly found herself pulled into a national debate.

After *Silent Spring* was published, President Kennedy and more than half a dozen senators and congressmen had launched investigations into pesticides. West flew to Washington to testify in the fall of 1962 and again in the summer of 1963. She told one subcommittee exactly what she had told Carson: parathion was a deadly poison, and it was being used in unimaginable amounts. The amount used on California farms alone, she noted, could decimate the world's population five times over. Thanks to uncontrolled use of such pesticides, she added, agriculture now had the highest occupational disease rate of any industry. It was the farm laborers who suffered most, given no way to protect themselves, not even washing facilities in the field. And their "migratory status" meant that many of their illnesses were never even shared with authorities. The pesticide poisonings her bureau had counted, she said, were most certainly a significant underestimate.

Meanwhile, back home in California, Sacramento lawmakers were also rushing to respond to the pesticide problem. Inspired by *Silent Spring*, State Senator Nicholas Petris of Oakland had introduced a bill to ban DDT. The bill sailed through the assembly, sponsored by Petris's colleague William Byron Rumford, of Berkeley, but a Senate subcommittee refused to let it move forward without an investigation. Rumford and Petris headed up to Davis, where the University of California's main agricultural campus was, to gather testimonies from scientists. Meanwhile, Governor Edmund G. Brown had created a special committee on pesticides. It also wanted West's expertise. Then UC Davis chancellor Emil Mrak, a respected food scientist, spoke out in defense of pesticides. Growers knew how to use them safely, he said. And their residues were not poisoning anyone. West, of course, knew they were. She realized she needed to find a new way to make their hazards even more clear.

The bracero program, West saw, was on its last breath. The House of Representatives had already voted to end it, and the Senate was likely to follow. Lest anyone think the poisoning problem would go away with it, she pointed out that three thousand California children had gotten emergency medical care for pesticide poisonings in 1960. Some were children laboring on farms. Others were middle-class casualties of a newly hazardous environment, and their stories sounded just like those that Lester Petrie had collected in Georgia. A toddler died after making mud pies with TEPP. A two-year-old died after eating an organophosphate fly cake. In an entirely different kind of case, a bale of blue jeans was shipped in a truck with a leaking drum of the pesticide phosdrin. The jeans were unloaded, packaged, and shipped to stores. Eight months later, the pesticide remained in the fabric. Parents who bought the jeans had no idea that they were dressing their children in clothing that would land them in the hospital.

•

Govea had realized her parents' dreams. She had graduated from high school and enrolled in Bakersfield Junior College, hoping to become a lawyer someday. She had just gotten home from class one afternoon when her father asked her to attend a meeting with Chavez in his place that evening. Neither of them had any idea that his request would change her life.

At the meeting, Chavez made an announcement. He had left the CSO to organize a new union for farmworkers; this, he told the group, was the change California needed most. Growing up working in the fields, he had seen farmworkers living in cardboard shacks, drinking from irrigation ditches, and subsisting on beans and dandelion greens. Now, in the 1960s, things were little better. Govea was overcome with anticipation as she listened to him. If there was going to be a union, she wanted to be a part of it. The National Farm Workers Association (NFWA) opened up a Bakersfield office. Chavez put Marshall Ganz, a recent Harvard grad from the white side of Bakersfield, in charge and then asked union families to volunteer. Govea jumped at the chance. She spent hours in the office teaching Ganz Spanish with flash cards. And the more time she spent there, the more she wanted to do.

Govea turned to her parents for their blessing, hoping they would let her put off college so she could work for the union full-time. Reluctantly, they agreed she could leave school for a year. In 1965, instead of enrolling in classes, Govea moved to Delano, thirty-five miles north, where the NFWA had established its headquarters. The union already had 1,500 dues-paying members. She woke early every morning to head out to the fields with other NFWA staff, armed with signs and chants. They urged grape pickers in the San Joaquin Valley to stop working in the fields until the growers honored the US Labor Department's minimum rate of $1.40 an hour. The NFWA paid her five dollars a week.

The strikes didn't work; there were too many nonunion workers willing to cross the picket line. But Chavez had other ideas. He convinced unionized dockworkers at the Oakland port not to load the fruit. A thousand ten-ton cases of grapes rotted on the docks. Then he called for a boycott of the two biggest grape growers in Delano, Cutty Sark whiskey maker Schenley Industries, with five thousand acres in Kern County, outside Delano, and DiGiorgio Corporation, with eleven thousand acres overseen by a wealthy San Francisco patriarch. The union held protests and picketed. Faith groups and liberal college students joined the cause. The United Auto Workers and the Teamsters voiced their support. Liquor sales in San Francisco and Los Angeles tumbled. The growers agreed to negotiate. Schenley signed a union contract with higher wages. DiGiorgio held out.

In neither case did the workers or growers discuss pesticides. The boycott continued, slowly bringing in other, smaller growers who gradually followed Schenley's lead.

In the meantime, Govea transferred over to working at the union's Farm Worker Service Center in Delano, where the union had its offices, credit union, and newspaper. Anyone could come in, for help of any kind. During the strikes the union had also set up a temporary medical clinic in a home in Delano. They hung a sheet over the kitchen doorway, hired a nurse, and brought in a rotation of volunteer doctors. The workers who came in shared their stories. Dehydration was common. Some foremen forced dozens of workers to share a single cup of water. Others charged the workers 25 cents to drink. Births were unattended, diabetes untreated. Emergencies were ignored, injuries neglected. The local hospital charged a cash deposit on admission, the county hospital was thirty-five miles away, and transportation was scarce.

When a couple who spoke no English came in one day, Govea offered to translate. Their toddler had just died. The little girl had rushed to hug her father when he had come home from work in the fields, burying her face in his clothes. His pants and shirt were covered with dust from a poison, sprayed earlier, that had shaken off the leaves as he worked. The dust filled the little girl's lungs, or it seeped through her skin. Maybe both. They didn't know. Either way, it killed her.

Govea remembered that dust. The union, she realized, had to do more about the problem of pesticides. Clinic staff agreed. The union created a health and safety committee to suggest clauses for union contracts and then reached out to the state health department to ask if it could survey farmworker health in the valley.

Chapter 14

A BAN

Irma West drove down through the East Bay hills to her office in downtown Berkeley, wondering—as always these days—if she was being followed. Someone from the union told her they had learned that a detective, probably hired by a pesticide manufacturer, was tracking her movements.

She wasn't terribly surprised. A few years earlier, an FBI agent had knocked on her door, showed his ID, and asked her to share some files she had. He somehow knew about her visit from a Swedish physician who had stopped by the Occupational Health Bureau on a trip to California not long before. The physician had given her papers on the toxicity of organophosphate poisons, such as TEPP and parathion, which she had been desperately trying to learn more about. He claimed he had them because Sweden, largely neutral during the war, had received them from the Nazis, who had tested the chemicals as nerve gases. When the FBI agent appeared, she gave him the papers—and never mentioned the copious notes she had taken.

As she approached the modern, eight-story building that housed her office, towering over the shops and restaurants that lined the streets on the edges of the University of California campus, she seemed to be alone. She had just come back from a meeting of the US-Mexico Border Public Health Association, where she had shared the results of her bureau's survey of conditions for farmworkers. The bureau had found the conditions abominable. Workers toiled without water for drinking, or washing, or toilets. Some soap and water alone, she said, would prevent some of the most serious health problems and even save lives.

But now she needed to prepare a talk for a very different audience. The American Chemical Society had invited her to Atlantic City to give a presentation that fall, the fall of 1965. In the wake of *Silent Spring*, the association was planning a symposium on the topic of organic pesticides in the environment, and it had asked her to come speak. She had given more talks than she could keep track of the last few years. The invitations had really poured in after she gave a talk about *Silent Spring* at the University of California, Davis. The book had merits and problems, but the agricultural college, supported by agricultural companies, didn't want to hear her opinions on its merits. The audience booed her off the stage. West had never been one to keep her strongly held opinions to herself, but as she considered what she wanted to say at the American Chemical Society meeting, she thought it might be time to take an even stronger stand than usual.

West considered the unique position she was in. Her bureau of five physicians, eight engineers, two nurses, and two statisticians relied heavily on the CDC's toxicology program for information on pesticide toxicity. Yet here she was, in a heavily agricultural state that used a fifth of the nation's pesticides, applying a hundred million pounds of the chemicals each year. Her bureau had decided to conduct its own studies because so many of the pesticide questions it faced simply hadn't been answered by anyone else. For instance, malathion was supposed to be extremely safe, but her bureau kept getting reports of skin disease among workers using it. Sure enough, when her own physicians tested it on prisoners, the prisoners broke out in painful, inflamed rashes. Organic phosphate poisonings, such as the one in Hughson, raised other questions. With atropine, the sickened workers recovered. But how often could a person suffer repeated severe poisoning without some other lasting effect? On top of everything else, as her bureau surveyed farmworkers, they learned of growers giving sick workers atropine right in the field and sending them straight back to work. At this point, she believed her bureau had created the nation's "only body of factual data" about the dangers that pesticides posed to people's health.

There were other things that bothered her about the national pesticide debate. Defenders claimed that pesticides were necessary to protect food crops and prevent disease. But now, in 1965, just 1 percent of all pesticides applied were sprayed to stop disease. The vast majority was sprayed in agriculture. And never before had hundreds of new chemicals with such varying degrees and potential for harm been introduced into the environment in such a short amount of time. She was surprised that so few physicians had spoken out about the problem. She could only assume that it was because they just weren't aware of it—or its scale. That was why she had also just given a talk to the Los Angeles County Medical Association. Little by little, she was trying to share with other physicians what she and her bureau knew.

As she sat down to write her Atlantic City talk, she kept thinking of two recent articles she had read by the microbiologist and author René Dubos. Dubos believed that scientists and the public had missed an opportunity to learn from their experiences with cigarettes and the type of ionizing radiation used in X-rays. Neither cigarettes nor X-rays caused anyone harm in the short term. But both had delayed effects that took years, or decades, to come to light. Yet today's scientists and physicians weren't developing new ways of thinking about or detecting the long-range effects of "modern ways of life." He didn't say pesticides, but he might as well have, West thought. There was another problem, too, wrote Dubos. Modern science was exceptional at producing "isolated fragments of knowledge," but it was far less successful in dealing with the complexity of living systems and the natural world. This, too, seemed so relevant to the pesticide problem. Scientists' traditional ways of thinking, West realized, were "actual barriers" to tackling the problems of the modern world.

In September, West flew to New Jersey with a call to arms on the typed pages in her briefcase. The symposium opened with a welcome from its organizers, Herman F. Kraybill and Aaron A. Rosen, both scientists for the US Public Health Service. Kraybill, a nutritionist, had studied food safety for the US military before moving to the environmental health division of the service in Washington, DC, to study pesticides. Rosen, a chemist stationed in Cincinnati, had been documenting water pollutants for the federal government since the late fifties. He gave a grim introduction. Pesticides were now everywhere, he reported: lakes, streams, rivers, birds, fish, forest animals, clothing, and of course people. Scientists had found them in the middle of the Pacific Ocean and in waterfowl in Antarctica, places where they had never been applied.

"When historians of future generations write of our era," he said, "they will note wryly how much of our technical capability was devoted to producing commodities inevitably destined to contaminate our environment."

Rosen and Kraybill had given the speakers a single directive: to discuss pesticides in the environment, not the food supply. The latter had already been exhaustively discussed, they said. West listened politely to the other presenters, then took the podium herself. She pulled out her typed pages. She had written a proposal: that scientists abandon toxicology as a way of thinking.

She began with an acknowledgment of the importance of pesticides. No one disputed that they had value, she said. But their use was predicated on three flawed scientific ideas. The first was the toxicologist's idea of dose. The second was toxicology's use of animal data. The third was toxicology's approach to scientific proof.

The toxicologist, said West, held fast to the notion of the threshold dose: that everything was safe at some level and hazardous only when a person or animal took in more than that dose. But that concept ignored the complexity of living systems and the ways in which chemicals interacted in living beings, she said. Two chemicals might produce a combined effect worse than either of them on their own. One chemical might quietly ruin the body's ability to detoxify another chemical. One chemical might amplify the effect of another chemical. The idea of a threshold dose ignored all of these possibilities.

The second problem, she went on, was toxicology's dependence on animal tests. Research had shown time and again that animal

151

tests were not a reliable indicator of human effects, yet toxicologists continued to depend on them heavily. Not only that, they tended to rely on those animals most convenient in lab settings—not those whose biochemistry most closely approximated that of people. As such, they were an inherently flawed means of creating scientific knowledge about toxicity.

Third, she argued, scientific norms held that a chemical was harmless until proven harmful. This, she said, placed the burden of proof on the public and the environment, protecting the economic interests that produced and profited from those chemicals. On this point, she decided to forgo scientific jargon. "Waiting for a sufficient number of dead bodies to provide the proof," she said, "is crassly immoral."

These problems, West told the audience, were destroying public trust in science. The attitudes of too many scientists further compounded the problems. Members of the public could not protest scientific decision making without being labeled "emotional." The public wanted facts, she said; they did not want to be "patted on the head periodically and told that the food supply, or whatever, is perfectly 'safe.'" They wanted scientists to listen to them, hear their concerns, and learn from them. Above all, they wanted scientists to "develop a greater sense of humility."

One place to start, she said, might be to admit that proceeding so far ahead of understanding, as the nation had so clearly done with pesticides, was "not in the best traditions of science." The audience clapped. And the conference moved on to other matters.

•

The pesticide cases were always obvious to Jessica Govea. Their foremen would often tell them that their dizziness, double vision, or vomiting was heatstroke. When they came into the service center and shared their stories, however, Govea knew that wasn't true. Some of the other volunteers who had worked in the fields also knew it. The question was what to do.

In the spring of 1967, Govea had been working in the service center about a year when Cesar Chavez offered her a new assignment.

The union, now called the United Farm Workers (UFW), had just hired its first general counsel to try to defend against growers using the legal system to block their pickets and topple their strikes. The lawyer, Jerry Cohen, was twenty-six years old and a year out of law school, and he needed an assistant. Govea, now nineteen, accepted the job and joined Cohen in the Legal Department: the kitchen of a little pink house next to the service center. What are we going to do about pesticides? she asked him right away. He said he didn't know. So she kept asking, over and over again.

Govea was still working in the Legal Department that fall when Chavez called an all-hands meeting. The union was striking at more than a dozen Delano vineyards, and the growers who hadn't yet signed union contracts seemed only more and more determined not to. So the union was going to expand the boycott to include not just wine but table grapes as well. When Chavez asked for volunteers willing to organize boycotts in cities across the country, Govea raised her hand. When he asked who was willing to go anywhere to organize a boycott, Govea kept her hand up. She had just turned twenty, and although she still hadn't gone back to college, her parents were proud of the work she was doing. She'd go anywhere.

Because Govea had kept her hand up, she was sent far: Toronto. Up until the day she left, she didn't know where it was. "I didn't know it was three thousand miles away," she said. "I didn't know it was in another country." She boarded a plane with Ganz and the Catholic priest who ministered full-time to Delano's farmworkers, Father Mark Day. When they landed, she found herself in a new world to do a job with no instructions: convince grocery store customers to stop buying California grapes until the growers used union labor. They had a single name to go on, and they spent their first day tracking him down: Dennis McDermott, Canadian director of the United Auto Workers. McDermott spread the word among the UAW and helped the trio set up picket lines throughout Toronto.

Govea quickly noted that although Bakersfield came in two colors—brown and white—Toronto came in one. Her first day out on the line, Govea looked at the sea of white faces and wondered how they would receive her. She saw two men walking toward her and steeled herself. They reminded her of white boys from Bakersfield. But she knew she had to talk to them. "Excuse me," she said, "could I ask you to help farm workers by not buying grapes?" In unison, they turned their backs to her. She took the move as a sign of rejection until she saw the UAW emblem on the backs of their jackets. "We're all for you," they said.

Toronto was the third-largest grape market in North America. Govea, Ganz, and Day spread the boycott one grocery chain at a time. When the Ontario Federation of Labor met for its annual convention, it asked the UFW team to come speak. Govea stood before the crowd of 1,200 white men and shared her story. The grapes of wrath still existed in California, she told them. Grapes were picked by families working twelve-hour days, in 115-degree heat, with no water, rest, or toilets. Growers got away with paying what they wished. The children were killed by tractors and spray rigs. "When you see those nice sweet grapes," she said, "we ask you to see the bitterness and sweat with which we picked them. We ask you to see the small children who lug grape boxes through the fields."

The speech brought the men to tears. The next day, it made the papers. The UFW expanded the boycott to Montreal, North America's fifth-largest grape market, and put Govea in charge. She was twenty-one.

Back in Delano, the summer of 1968 saw a rash of pesticide poisonings. Dozens of farmworkers fell ill in June and July, in one case while working in a vineyard that had reportedly been sprayed more than a month before. Then a trio of workers came into headquarters with nausea, dizziness, and blurry vision, sick because they had been in the fields when the spraying happened.

Cohen, the UFW attorney, thought of Govea's persistent question. Growers were supposed to report all spraying to the Kern County agricultural commissioner, so he went straight to the commissioner's office and asked to see the records of what had been sprayed. The commissioner told him to come back the next day. But when Cohen returned, the commissioner told him that the records were off-limits. The crop-dusting company responsible for the spraying had gotten a restraining order: neither he nor any union member could view the records—or any spray records, for that matter. And then the Kern County Agricultural Chemical Association, which represented all of the local crop dusters, obtained a court order to keep the records sealed permanently, claiming that they contained trade secrets.

The union told West's bureau up in Berkeley. West stepped in herself by going straight to the press. State law required a public record of all pesticides used on farms, she told reporters. Her colleague Tom Milby, who had responded to the peach-picker poisoning in Stanislaus, released the bureau's data on pesticide poisonings for the year. Close to a hundred pesticide poisonings had happened in Kern County alone. And that was only counting the ones that they knew of.

Cohen knew his next move. He filed suit against the commissioner. Then he filed a separate suit on behalf of a seventeen-year-old boy in Tulare County who had become so sick after exposure to an organophosphate pesticide in an orchard that he could not work for the next five months. "We are determined to make the pesticide issue a new battle line," he told the press.

Up in Montreal, Govea was spending the winter in a place colder than anywhere she had ever been. With the support of the UAW and the Canadian Council of Churches, she took the boycott from one grocery store to another, holding vigils outside in the bitter chill and snow, handing out leaflets to one customer at a time. It was grueling. She was so short on cash that there were days she ate nothing more than a candy bar. She was so far from home, and her family, that she often cried herself to sleep at night. But the boycott was working. Not just customers but whole grocery chains were agreeing to stop buying nonunion grapes. So she kept at it.

West and Milby had tallied the state's number of pesticide poisonings at the UFW's request, but they had quickly come to realize that the numbers were complicated. More than half of the hundred people poisoned in Kern County were farmworkers. Most of the rest were those applying the pesticides themselves, including the crop-dusting pilots. But there were also those made sick in tragic accidents. The pattern was the same in county after county. In one case, bakery workers sprayed for pests, not realizing they had contaminated their own flour until twenty-eight customers fell ill after eating their doughnuts. Farmworkers were most of the picture but not all of it.

On top of that, pesticide deaths and illnesses fluctuated, up one year, down the next, higher than ever the year after that. They were happening everywhere, too, not just on farms. Five years earlier, West had warned rural doctors to look out for pesticide poisonings. Now she issued a new report, telling doctors throughout the state to know the signs. Her bureau had also noticed something else. Most of California's poisoning cases could be attributed to three pesticides: parathion, phosdrin, and thimet. All three were organophosphates—the same class of chemicals discussed in the papers seized from West by the FBI—whose use generally was on a sharp rise.

Yet all anyone seemed to be talking about were the chlorinated hydrocarbons, especially DDT. In the summer of 1968, as organophosphate poisonings climbed in California, the FDA lowered DDT tolerances on thirty crops. The new tolerance for milk was 0.05 ppm; for butter, it was 1.25. When a shipment of 12,000 pounds of Arizona butter was seized by federal officials in Los Angeles, Arizona's dairy industry called for a ban, and the state now had a DDT moratorium in place. Michigan and Wisconsin were close to doing the same. The tide was clearly turning against DDT—which was a good thing—but it wasn't the pesticide killing people in the fields. West was once again fighting a battle in which no one else seemed terribly interested—except the UFW. "Thank you for your generous contribution of love/service and $5.00" said the notes of gratitude she received, signed by Chavez, each time she sent the union another check.

•

In January 1969, Chavez wrote to three organizations of table-grape growers: the Grape and Tree Fruit League of San Francisco, the Desert Grape League down in Mecca, and the South Central

Farmers Committee of Delano. When they didn't reply, he called a press conference in Delano, and the big news wires showed up. The union had noticed a growing number of pesticide illnesses in its clinic, he told reporters. They had asked growers to talk about the pesticide issue—even if it meant putting talks on wages and working conditions aside for the moment. And because the growers weren't responding, the UFW was going to take it on itself to tell the public about pesticide dangers. But Chavez carefully avoided use of the word *pesticide*.

"We will be damned," he said, "if we permit human beings to sustain permanent damage to their health from economic poisons."

The UFW had made a quick and calculated shift, putting pesticide poisoning front and center in its campaigns for 1969. It made plans to ask for a pesticide clause in union contracts and began to incorporate the dangers of pesticides in its messages to union members and the public. "This will be for the protection of both the workers and the consumers," a union staffer said.

The union's leaflets, like the ones Govea and her boycotters handed out in Montreal, told harrowing tales of the dangers of sprayed grapes. A picker described vomiting and bleeding from the nose after eating grapes straight off the vine. Another suffered vomiting, sleeplessness, and difficulty breathing and seeing for ten days after getting caught in a spray rig's path. The poisons came in two kinds, the leaflets said. There were the "nerve gases developed by the Nazis" and the "hard" pesticides, which included such "long-life pesticides" as DDT. "We continue to be the guinea pigs in a base poison experiment motivated by profit," they said.

The union sometimes deliberately conflated the risks of the organophosphates with DDT. Hundreds of workers were poisoned every year, and thousands more suffered skin infections and burns from pesticides, the union told the public. And then they added what they knew to be true about DDT: more than 100 tons were sprayed annually over vineyards. It was in nearly all American foods and every American body. Studies linked it to cancer in lab animals. Levels in farmworker children were "frightening," and their rate of birth defects and cancer was "staggering." The only reason that DDT was still on the market, the union charged, was because of growers' close relationships with political figures in Washington. Late in the spring of 1969, the union called on supermarket chains to test their grapes for DDT and pull them from the market if they found it. If they didn't test, the union was going to keep telling the public about DDT's dangers until they did.

Not everyone agreed that DDT should be part of the union's message. Parathion should be outlawed first, said one union member. The newer, more poisonous pesticides coming into use deserved more attention, said another. The union held firm. DDT might not be visibly poisoning workers the way parathion was, but it was just as widely sprayed on grapes, the UFW's newspaper, *El Malcriado*, reported. And farmworkers had DDT levels three times higher than those of other Americans. Because most of the women breast-fed their babies, they were spreading whatever still-unknown hazard it carried to their families at a higher rate, too. "Time may be running out for us and our children," *El Malcriado* stated.

In the Kern County court that fall, Cohen presented one farmworker after another who shared their stories of pesticide illnesses. Men who worked in the orange groves said they were told to quit when they complained that the work was giving them rashes. Men who picked in grape fields said their fingernails turned black and fell off. A man working on a ranch became sleepless, sightless, and nearly unconscious after spraying vines with parathion. A man who fell sick after being directly sprayed was forced by his employer to pay for his own medical care. A woman said that when she asked a ranch owner why he sprayed so close to the workers, he said it was to make the women's breasts bigger. And then he laughed. As the stories filtered out of the courtroom, local growers fumed.

In the meantime, a small group of table-grape growers, going bankrupt from the boycott, had agreed to meet secretly with the union. Union representatives tried to convince them that the union label was all it would take to win back consumers. The growers were skeptical. The 1969 grape crop had already been sprayed, back in the spring. Hearings in California and in Washington, not to mention the union's own messaging and news reports on both, had already informed the public that grapes were tainted with "long-life" pesticides. The talks broke down.

But in Kern County, one wine-grape grower, Perelli-Minetti and Sons Vineyards, stayed at the table. While Cohen extracted testimonies in court that September, Perelli-Minetti agreed to sign a union contract providing vacation pay, benefits, equal-opportunity hiring, and a health and safety clause. The clause banned the use of DDT, along with aldrin, dieldrin, and endrin. It was the nation's first ban on the "hard" pesticides, with DDT topping the list. "Company and union agree," the pioneering Perelli-Minetti contract said, "that the subject of economic poisons is a necessary and desirable subject for this collective bargaining agreement."

"Hey, you other growers," *El Malcriado* said, "it's time to sign."

It took the better part of a year, but they did. In the spring of 1970 a table-grape grower in Coachella signed with the union. That July, Giumarra, one of the biggest table-grape growers in the world, did too. After Giumarra, twenty-six growers representing 70 percent of the US grape crop signed. By the end of the season, the union had signed 150 table-grape contracts covering 30,000 workers. The contracts raised wages, aided displaced workers, put employer contributions into a health and welfare fund, and banned parathion and TEPP, in addition to the hard pesticides listed in Perelli-Minetti's contract. They also gave the union the right to organize a health committee with access to all spraying records.

In Bakersfield, Chavez went to Ganz's office in the little pink house to call boycotters who were far from home to let them know. He called Govea first. "Oh wow. This is like heaven," Govea said. "I'm not going to be able to sleep tonight!"

El Malcriado published a profile of her. She was twenty-three and credited with running a boycott that had cut grape shipments in half. "I no longer belong to myself," she said, "but to the thousands of people who are struggling to be free . . . farm workers, native Indians, blacks, Vietnamese."

Wrapped up in the first big victory for unionized farmworkers was the beginning of the end of DDT. California followed the union's lead, banning DDT on grapes and dozens of other crops in the spring of 1970. And there was such an uproar over the chemical nationally that it actually seemed as if the FDA itself might ban it from food crops entirely.

UFW staff members were circumspect. Said an editorial in *El Malcriado*, "A ban on DDT, while definitely a step in the right direction, would only scratch the surface of the problem."

Chapter 15

THE BIRDS

Fred and Marguerite Baumgartner courted on bright moonlit nights spent listening for the hoots of great horned owls in upstate New York. When they finished their degrees, they moved to Oklahoma. Fred studied bobwhite quail and grebes. Marguerite was an expert in the tree sparrow, whose song, she thought, sounded like it was calling her name. They loved birds, and because they loved birds, they loved bird habitats. So when they moved to Wisconsin in 1965, for Fred's new job at the state university in Stevens Point, they joined a group called the Conservation Natural Resources Association, whose members met in people's homes to talk about how to protect natural environments under threat.

The CNRA wasn't a big organization: some two hundred people—academics, farmers, teachers, and nature lovers—scattered across the state. But together they pressed the highway department to plant "native" trees and shrubs instead of "exotic" ones. They challenged an Army Corps of Engineers project on the Wisconsin River. And after holding a conference on pesticides in 1966, they began discussing what they should do about DDT. The Wisconsin Conservation Department had just banned its staff from using the pesticide. Its director was asking everyone in Wisconsin to stop using it near lakes and streams. One of the department's scientists explained why: "Every fish we've checked has DDT."

DDT was always prized for its persistence, but that chemical characteristic had implications that were becoming clear to scientists who studied fish and birds, such as the Baumgartners. DDT's persistence meant that fifteen or maybe even twenty years after it was sprayed, only half of it had disintegrated or decayed; the other half was still in nature. But even as the amount of DDT in soil or water slowly diminished, over time the amounts in certain forms of wildlife were increasing. It all had to do with DDT's predilection for fat.

When small fish fed on plankton, they excreted the waste but held onto the DDT in their fat tissue. With each feeding, the small fish stored up more and more DDT. When big fish fed on those smaller fish, they stored all of the DDT from those fish, which gave them higher concentrations of DDT in their own fat than the little fish had. The same was true of birds of prey. Animals at the top of a food chain, in short, consumed and stored the DDT of every organism below them. This meant that in some of those animals, DDT could reach fatal levels even if none of the organisms below them on the chain had suffered.

University of California, Davis, zoologist Robert Rudd, who was studying the phenomenon closely in birds, called it "delayed expression." It was an "insidious" process, he said. "It goes unnoticed until mortality results, and even then the cause of mortality may often not be suspected." The effect was starkly demonstrated by wildlife biologists investigating the infamous collapse of a population of western grebes at Clear Lake, California. Back in 1948 the lake was home to a thousand mating pairs of grebes, long-billed birds known for a graceful courting dance in which they skimmed across the water, wings wide. A year later, Clear Lake's grebes were gone. Over the next ten years, scientists would occasionally see a handful of grebes, and then they too would die off. Biologists who showed up at the lake in 1957 examined its dead birds for disease; finding nothing, they decided to test the birds for DDD. The local mosquito commission had sprayed the lake with the insecticide—which differed from DDT by a single molecule of chlorine—three times in the 1940s and 1950s to kill gnats. They used a solution calculated to result in 14 parts per billion DDD in the lake, enough to kill gnat larvae and nothing else. Or so they thought. It now appeared they'd been wrong. To the biologists' shock, the dead birds contained 1,600 ppm DDD in their fat, eighty thousand times the concentration applied to the lake.

When Rudd learned about the Clear Lake study, he knew its implications likely meant widespread devastation for birds. As an aside, he also worried what this delayed expression meant for people who ate fish. As it happened, his own term for the phenomenon—delayed expression—did not persist. Scientists would later call it biomagnification, the end result of the bioaccumulation of persistent chemicals in individual organisms in a food chain.

When Joe Hickey saw the report on the Clear Lake grebes, it troubled him too. It also seemed familiar. Hickey, a beloved wildlife professor at the University of Wisconsin, Madison, had fallen in love with birding as a boy growing up in the Bronx. In the late fifties, he had also carried out a study on DDT's effects on birds. At the time, the state's agriculture department was urging cities and towns to spray DDT to kill a bark beetle that was killing elm trees. The beetles spread a fungus that clogged the tree's vascular system, causing a blight known as Dutch elm disease. Hickey and a colleague had decided to compare the numbers of birds in three communities that sprayed and three that hadn't. Several species declined in the sprayed areas, but none more than the robin, whose numbers plummeted. They then compared bird populations on their own university campus before and after it sprayed its 700-plus elms. Yellow-bellied sapsuckers died a few days after the spraying. Robins started dying after a week. Starlings and grackles followed. Piecing together his own study and that of the grebes, Hickey realized that there were two ways persistent pesticides killed birds: right away, over days and weeks, and slowly, over years.

The accumulating research on birds and chlorinated pesticides was enough to convince the Wisconsin Conservation Department to curtail its DDT use in the early sixties. But in the late sixties, other state agencies, especially the agriculture department, were still using it—largely because Dutch elm disease was on the rise. In Wisconsin cities like Madison, where streets were lined with the tall and stately trees, the disease was killing more elms every year. DDT, however, seemed to stop its spread. So Janesville sprayed it. As did Green Bay. And Waukesha. The Milwaukee County town of Wauwatosa had the oldest and tallest elm in the state, a 296-year-old specimen more than 21 feet around at the ground. In 1968 it was so badly diseased that the town cut it down and sent out a team to spray.

City arborist Allen Prill was following his supervisor's orders, spraying on a breezy day, when the winds shifted and the DDT blew back on him. "It was bitter," Prill said. He noticed birds drinking from puddles in the street below him beginning to tremor. He continued to spray.

Over in the Milwaukee suburb of Bayside, a CNRA member named Lorrie Otto was leading Girl Scout troops through the woods near her home when she too noticed birds convulsing. Otto was an ebullient nature lover with a fondness for bright outfits. Her front lawn was a sea of wildflowers that irked her neighbors. She told her fellow CNRA members she couldn't finish *Silent Spring* because she found it too depressing. She was convinced that the birds in her town were suffering because Bayside had sprayed its trees with DDT. She began collecting dead birds as she found them, storing them in her freezer in case she needed to prove it someday.

Otto, like Baumgartner, had attended CNRA's pesticide conference in 1966. She remembered the University of Wisconsin ornithologist who spoke: Joe Hickey, who had described the birds on his campus killed by DDT. She reached out to ask if he thought that was what was happening in Bayside, too. Yes, Hickey told her, it sounded like DDT to him. But he was reluctant to do anything about it. As much as bird declines grieved him, he said he saw himself as a scientist, not an activist.

Otto, dejected, complained to a friend. The friend told her about a professor in New York who had also studied DDT's effects on birds and who was part of a local organization, like hers, that was suing local governments over their spray plans—and not just in New

York. Right next door in Michigan, the professor's organization had sued the Michigan Department of Agriculture over its use of dieldrin and, separately, fifty-two cities for using DDT. The organization got an injunction in the first case, a series of court orders in the second, and a lot of publicity for both, which were still ongoing.

Otto didn't want to go to court, but she wrote to the professor, a biologist named Charles Wurster, anyway. She thought his organization, the Environmental Defense Fund (EDF), might be willing to help CNRA get a story about the DDT problem on television or the radio so that more people in Wisconsin could learn what she, the rest of her organization, and the state's bird experts knew.

But when Otto shared her story and her request with Wurster, he declined—and then tried to persuade her of a different course of action. As a professor, Wurster told Otto, he valued and believed in education. Yet when it came to the matter of pesticides' harm to wildlife, a subject scientists had known about for decades, education was just too slow. His organization, he said, had accomplished more in a few months than education could accomplish in five lifetimes. It was an organization for people who had lost patience. It was for those ready to take legal action. He'd be happy to help the CNRA members, he said, as long as they agreed that "the way to stop the DDT in Wisconsin is either to threaten to take the offenders to court—or to actually take them to court."

Otto was convinced. She brought the proposal to Baumgartner; the mild-mannered professor and longtime Boy Scout leader was now also president of the CNRA. Wurster's organization, she told him, had been nothing but successful so far, even if it had been around for only a few months. Baumgartner, a birder out of patience himself, agreed to get the organization behind the suit.

At the annual CNRA meeting in the fall of 1968, in the little town of Spring Green on the Wisconsin River, Baumgartner told members that he believed they had a chance to stop the use of DDT in their state. He said he would be personally filing a complaint with the city of Milwaukee, charging that runoff from any further DDT spraying would kill birds and pollute Lake Michigan. But if the CNRA could raise $25,000, he said, lawyers with the EDF in New York would help them tackle the problem at the state level by suing the Wisconsin Department of Agriculture. Please, he said, write checks, write letters to the governor, and get ready for a battle against DDT. The members signed on.

•

Tall and trim Charlie Wurster had become the person to call inside of four short years. It had all started when he was at a party in Hanover, New Hampshire, with his then-wife Doris, in December 1962. A local birder at the party had circulated a petition to stop the town from spraying its elms with DDT. The petition didn't work, but it did get Wurster and Doris thinking.

The two were newly minted PhDs from Stanford, he in chemistry and she in biology, and he had always been interested in birds. Wurster hadn't read *Silent Spring*, but he and Doris had both noticed a lot of talk about birds and pesticides of late. As scientists, they decided to carry out a small study: they'd simply count birds before and after the spray, both in Hanover and in Norwich, Vermont, a town across the Connecticut River with no spray plans. The city planned to spray in April, so early that spring they ran ads in the local papers asking people to bring sick and dying birds to their offices on the Dartmouth College campus.

A student brought in the first specimen at the end of the month. Wurster took the twitching robin into his hands. Within minutes, it was dead. The robin was the first of more than 150 dead birds that he and Doris collected in Hanover that spring. In Norwich, they found ten. By June, they estimated, Hanover's robin population had declined 70 percent. The city's chickadee, nuthatch, creeper, and woodpecker populations also fell. As the Wursters pored through wildlife journals, they realized that scientists before them, such as Joe Hickey in Wisconsin, had documented similar declines.

The Wursters sent some of the birds to a lab for analysis. Birds that had been found tremoring before death had more than 50 ppm DDT in their brains. Chipping sparrows, warblers, and white-breasted nuthatches had the next-highest DDT levels in their bodies. By then, Hanover had been spraying DDT annually for fifteen years.

The Wursters published their results, Charlie taking the lead on the version they sent to *Science* (the nation's top scientific publication), Doris taking the lead on the version they sent to *Ecology*. Then they divorced. Doris stayed in New Hampshire. Charlie moved to New York for his first job as a professor, in the biology department at Stony Brook University in Long Island's Suffolk County.

As Wurster settled in, a colleague from campus invited him to a meeting of the Brookhaven Town Natural Resources Committee (BTNRC). The group of mostly men—professors like him, high school teachers, and scientists at the nearby Brookhaven National Lab—met in one another's homes to talk about pollution and habitat destruction over coffee and doughnuts. Wurster thrived in their mix; the conversations felt vital, and the men were good company.

At a meeting in April 1966, the group's conversation turned to DDT. The Suffolk County Mosquito Commission was spraying it on local marshes. The commissioner was such a staunch DDT defender that, it was said, he liked to put a pinch of it in his mouth to disarm its opponents. The men in the group found this antic insufferable. They also knew all about Wurster's paper in *Science*, so they pressed him to write a letter of protest to the editor of the *Long Island Press*. Wurster agreed. They shared facts, figures, and anecdotes about declines in local fish and bird populations, and Wurster, never one to strain for words, got to work. "If the decline in Long Island wildlife is to be checked, the use of DDT for mosquito control must be curtailed," he wrote. "It is alarming to think that the dissemination of such toxic materials is in the hands of a person who thinks they are harmless."

Wurster's letter ran in the *Press* on a Friday in May. That night, his phone rang. The caller was a thirty-one-year-old lawyer named Victor Yannacone who told Wurster that he was representing his own wife, Carol, in a suit against a mosquito commission for flushing its DDT spray tanks into Long Island's Yaphank Lake, killing its fish. Carol, he said, had loved the lake as a child. He asked if Wurster would support the suit. Without hesitation, Wurster agreed. He recruited a few other BTNRC members, and they met with Yannacone the very next night. They talked and strategized, made copies of their scientific publications, went home to write affidavits on DDT's harm to nature, and sent the stacks of papers off to Yannacone to see what he could do.

Yannacone was a quick-thinking and often brash Long Islander who, in his legal work, took inspiration from the Legal Defense Fund of the National Association for the Advancement of Colored People and the American Civil Liberties Union; he had worked pro bono for the NAACP right out of law school. As he saw it, those organizations had broadened the definition of constitutionally protected rights. Following their lead, he planned to argue that a clean environment was every Long Islander's right, too. He also argued that because DDT carried a skull and crossbones on its packaging, it should be used with far more care than the commission had shown. The plan seemed to work. By summer's end, a local judge had issued a temporary injunction against the commission and given Yannacone a court date. He was going to trial, with the BTNRC men as his expert witnesses behind the scenes.

In the meantime, local papers covered the injunction. When the commission sprayed anyway, Yannacone claimed contempt of court, and the papers covered that, too. Ultimately, the judge ruled that DDT was causing damage to the Suffolk County ecosystem—but he also determined that only the legislature could stop the commission from using DDT. By then, however, the negative publicity and public protest had caused so many headaches for the commission that it simply dropped the chemical. If the agency needed to kill mosquitoes, the commissioner said, they would use the new pesticide Sevin instead.

"We had won while losing," Wurster marveled. He was convinced that Yannacone was on to something.

In the fall of 1967, Yannacone, Wurster, and a few of the other BTNRC members decided to incorporate as a new, more ambitious organization that, Wurster explained, would "marry science and the law to defend the environment in the courts." It was a grand ambition given the team's fairly limited record. And Wurster had had dinner with renowned ornithologist Robert Cushman Murphy, who lived nearby, enough times to know that others—like Murphy and Marjorie Spock—had tried just that in the past and failed.

But Wurster believed his group's strategy was different. They had the support of the well-heeled National Audubon Society behind them because in college Yannacone had worked for a professor who was on the society's board. And they knew that even if they didn't win in court, they could use the legal system to rev up local reporters and residents, and force government agencies to change course. They could keep on winning by losing. On the suggestion of an Audubon staffer, they called themselves the Environmental Defense Fund (EDF). Yannacone filed the requisite paperwork, and they incorporated as a nonprofit organization in the state of New York, with nascent hopes of someday achieving national reach.

They didn't have to wait long. Yannacone's former professor was H. Lewis Batts, who taught in the Biology Department at Kalamazoo College in Michigan. Batts had an idea for their next case, and he knew of a foundation that would grant them their first funds: $10,000 to stop the Michigan Department of Agriculture from spraying dieldrin. Less than two weeks after the EDF incorporated, Wurster and Yannacone were in Michigan, staying up all night to draft their first out-of-state complaint. And they were still tied up with the case in Michigan, which had snowballed into two cases and a lot of press by early 1968, when Wurster got a curious piece of mail at his office back in New York. It was a letter from a friend of a friend named Lorrie Otto, asking if he could help her stop her own state, Wisconsin, from spraying DDT.

The EDF was just a few months old. Neither Wurster nor Yannacone knew much about Wisconsin, but they did know a few things. They knew it had a lot of conservationists, people like them. They knew it was a big dairy state, and they knew that dairy interests in other states, such as Arizona, were turning against DDT. They also knew that a DDT case would give them the best chance of presenting irrefutable scientific evidence in court, because Wurster couldn't imagine a single wildlife biologist claiming it was harmless to fish or birds. If they took on a case in Wisconsin, thought Wurster, they just might win. He wrote back to Otto to see if she might be interested in seeing the issue go to court. She was.

On October 2, 1968, the EDF filed suit in federal court in Milwaukee, seeking to block the Wisconsin Department of Agriculture from spraying DDT on the state's elm trees. And then—as on Long Island and in Michigan—something they hadn't planned for happened. Maurice Van Susteren, a longtime hearings examiner for the Wisconsin Department of Natural Resources, approached them with a deal. Milwaukee wouldn't spray, he offered, if the EDF promised not to sue.

But Van Susteren also wanted them to know that they had another option, one that might stop DDT's use at an even higher level. A Wisconsin statute allowed private citizens to petition a state agency for hearings on a question related to the agency's authority, and by law, the agency had to respond. If private citizens asked the Department of Natural Resources to determine whether or not DDT was a water pollutant in Wisconsin, Van Susteren explained, the department would have to hold hearings and come to a decision. If the department found that DDT qualified as a pollutant according to the state's water-quality standards, it could ban any use of the chemical that resulted in water pollution—which would apply to just about any use at all.

Wurster and Yannacone went back to Baumgartner and Otto's group and filled them in. It seemed like an option worth pursuing, but the CNRA members wanted to make sure they had a backup plan in case it didn't work. So Baumgartner filed an official complaint with the natural resources department. DDT sprayed in communities along Lake Michigan, he claimed, was running from the streets into sewers and from there into the lake, where it had already killed a million coho salmon fry and vast numbers of uncounted birds. As another backup, he sought an injunction from the state's attorney general. The attorney general's office declined—only the governor, legislature, or a state agency could request an injunction, it replied. In the meantime, the EDF asked the Department of

Natural Resources to evaluate whether DDT was a water pollutant, and the department concurred. It set a hearing date for December 2, 1968.

•

The hearing date catapulted Wurster into a frenzy of activity. Now he was teaching classes, running experiments in his lab, and recruiting scientists to testify and provide statements in Wisconsin. But in no time at all, he had more witnesses than the EDF knew what to do with. Everyone he approached said yes. When Yannacone heard that a Swedish government committee had just recommended a national ban on DDT in that country, he told Wurster to call Sweden to ask the committee chair, a scientist named Goran Lofroth, to testify. It was a long shot. Otto had raised thousands of dollars from wealthy Wisconsin garden clubs, but the EDF was still operating on a shoestring. When Wurster reached Lofroth, he told the Swede they could cover his airfare but nothing more. It didn't matter. Within two days, Lofroth was in Madison. Otto put him up in her house.

Even if she hadn't, another volunteer would have taken him in. The CNRA, thought Wurster, was a "well-organized small army." They gave witnesses lodging and meals, raised funds, set up typing pools and a messenger service, and coordinated a team of University of Wisconsin students to do research. Anytime they needed something looked up, a student ran to the library. They were always back within the hour.

When the hearings finally got underway, they very quickly became a bigger event than the EDF or the CNRA could ever have anticipated. Part of the draw was Yannacone, a boisterous and booming courtroom personality. Otto called his courtroom style a "masterpiece." Wurster had developed an enormous respect for him; he grasped scientific material on a dime, cross-examined like a bulldog, and shot holes through the testimonies of witnesses for the agricultural chemical industry. Yannacone's first move in the hearings was to call for questioning Louis A. McLean, an attorney hired by the NACA, an industry group representing all pesticide makers.

Yannacone fired off questions about comments McLean had published, describing pesticide opponents as "quacks" obsessed with "sexual potency." McLean, chagrined, withered. The association pulled him. Yannacone had made it clear he was a force to be reckoned with. "Aim him in the right direction," said Wurster, proudly. "He'll raise hell."

As the hearings continued, reporters jammed the State Capitol building, vying for a glimpse of the action inside. They prowled the Madison airport, waiting to see who was coming to town to testify next. Students held demonstrations outside, calling themselves the DDT Commandos. They squirted spray guns at the trees. They carried signs that said "Liberate the Ecosystem" and "Ban the Bug Bomb." Otto found them charming. They were mostly science majors, she noted, some "no-nonsense" and others "gaily dressed."

In preparing for the hearings, Wurster and his EDF colleagues hadn't just come up with a list of experts; they had also devised a strategy. Previous suits, such as Murphy and Spock's, had tried to win against DDT by pointing to its known harms. But the EDF members decided to emphasize that environmental science, as they called it, was so necessarily interdisciplinary and inherently complex that no single scientist could possibly claim to understand it on his or her own. The same, they argued, was true for understanding the harms of a persistent chemical like DDT. It took a network of scientists to piece together its effects on intricate ecosystems and to understand how it was transported in water or soil, taken up by plankton or raptors, or metabolized by every species that encountered it.

So they selected scientists who could describe DDT's effects in nature in as many ways as they could conceive. And when they selected a scientist to testify, they ran that scientist's papers by a volunteer team of other scientists, including mathematicians, ecologists, and engineers, to try to catch any weaknesses or flaws. Then they sat the witness down with Yannacone to see how they held up. In the end, their witness list was a group of scientists who described DDT's effects on everything from local crustaceans to penguins in Antarctica—where it hadn't even been sprayed.

The goal there, Wurster said, was to prove that DDT was "wholly uncontrollable." The other goal was to focus, as much as possible, on birds. To that end, they invited Joe Hickey, the ornithology professor who eschewed activism, to take the stand. This time, Hickey agreed.

In his Bronx accent, Hickey told the story of an international conference on the peregrine falcon he had organized in Madison in 1965. The world's experts on the predatory bird had all noticed that the species was in sharp and startling decline; at the conference, they shared their respective observations. Across the globe, they all reported, the birds' eggs were either breaking or failing to hatch. The conference, in turn, inspired many of them to look more closely at the problem once they returned back home. An ornithologist in England did so by comparing modern-day falcon shells to those of blown eggs in British museum collections going back to the 1890s. The modern shells were markedly lighter and thinner. The same was true of golden eagle and sparrow hawk shells. Hickey himself worked with a graduate student to compare 34,000 eggshells in North American museums with the shells of modern falcons, bald eagles, and osprey. The modern shells were 19 percent thinner than those laid before 1947.

Hickey paused in his testimony to show an image. It was a photo of a bald eagle's nest, taken during egg collection. A single, wide-eyed chick peered up at the camera. Its wing gestured at a broken egg next to it in the nest. Thin shells broke under the weight of adult birds. That's why bird populations were declining.

Hickey was one of several scientists who had gradually arrived at a fairly consistent theory, drawing on findings from a handful of scientific fields. Pharmaceutical biochemists at a lab in New York had noticed that DDT and other organochlorine pesticides, such as chlordane, triggered the synthesis of liver enzymes capable of metabolizing drugs. In lab rats, the triggering effect was so pronounced that rodents exposed to the pesticide barely felt the effects of the powerful sedative phenobarbital; the elevated liver enzymes quickly broke down the drug. The same liver enzyme, it turned out, also broke down sex hormones, including estrogen. In female birds, estrogen signaled the body to store calcium in hollows in the bones, especially the femur, where the birds stored it until they needed it to form eggshells. Hickey and others hypothesized that by breaking down estrogen, DDT was disrupting birds' ability to metabolize calcium, leading to thin and fragile eggs.

But it was just a theory. The hypothesis was proven only when a team of scientists led by Lucille Stickel, a wildlife biologist at the Department of Interior, fed DDT to American kestrels, close relatives of the peregrine falcon, in a controlled experiment. On the DDT diet, the thickness of their eggshells declined 10 percent. Stickel's team also fed DDE, a metabolite of DDT, to mallard ducks. Their eggshells were 13 percent thinner than normal, cracked six times as often, and produced as much as half as many healthy hatchlings in the fourth week of incubation. Her scientists were just finishing the experiments while Hickey was testifying, but she agreed to board a last-minute flight to Madison to share the results.

Otto listened to Stickel's testimony, taking notes as she carefully spoke. Stickel was the only female witness. She was also the only witness to refuse to let Yannacone see her testimony beforehand. But Otto could tell, by the way the room received her, that her words were sealing the case against DDT. "You should have seen the men queen her!" Otto wrote to a friend.

That night and every night after the hearings adjourned, Wurster and Yannacone invited the reporters to have dinner with them. One of them, a writer who covered the latest scientific news for *Science*, asked if his journal could put Hickey's photo of the baby eaglet on its cover. Two months later, the fragile little bird peered up at subscribers and readers all across the country.

Baumgartner, meanwhile, spoke to the Wisconsin Assembly in favor of a bill that two lawmakers had just introduced to ban DDT. He stepped up his behind-the-scenes efforts to fund the hearings, driving all over the state to collect donations and deliver them to the EDF. Privately, he worried that the CNRA would be dragged down, accused of lobbying when it wasn't authorized to do so. But he also watched in sheer awe as the Madison hearings appeared not only in local papers but also in national papers and magazines, and as they stretched out over months. He sent Wurster a letter thanking him for what the EDF had done for their cause. "I can assure you," Baumgartner wrote, "that a small group of us in Wisconsin have given you a high place in the history of man's efforts to adjust and improve his environment."

Chapter 16

TOBACCO

On a Saturday in January 1964, Surgeon General Luther Terry, an Alabama physician with a high forehead and a smoking habit, stood at a lectern before a roomful of reporters in Washington, DC. A long line of white men sat in a row of chairs behind him. The men were experts the surgeon general had tapped to evaluate the evidence on the relationship between smoking and health, which had been accumulating in recent years. The task, which involved reviewing some 7,000 scientific papers, had taken the men and 150 assistants well over a year. Their report, *Smoking and Health*, summarized their findings: smoking cigarettes substantially increased a person's risk of chronic bronchitis, emphysema, coronary heart disease, and—most of all—lung cancer. In short, Terry told the reporters, smoking increased the risk of death. The news, he said, "hit the country like a bombshell."

He had known that it would. More than 40 percent of Americans were smokers. He himself had given up cigarettes, but he couldn't do without his pipe. The tobacco industry was such a massive economic driver that his office had scheduled the press conference for a Saturday to minimize its effects on the stock market—and to maximize publicity: the next day, the report was headline news all over the country and even in other parts of the world.

Some who heard the news thought they knew the source of the problem: they imagined that the tobacco companies had no idea how their product was grown. Down in Georgia, Mamie Plyler had watched for two decades as the tobacco fields adjacent to her farm were sprayed more heavily with each passing year. She knew that tobacco growers were often far removed from their own land: their crops were planted, sprayed, and harvested by field hands, then bought by dealers who sold the leaves at auctions held across the tobacco-growing states. In the wake of the surgeon general's report, she decided to write to the tobacco company executives herself because she knew they were even farther removed than growers from the fields she knew so well. "Have you made any research studies on why cigarette smoking is being so closely linked with these various diseases?" she wrote. She suggested they unite in an effort to study the chemicals applied to tobacco and the soils in which it was grown. "I am confident you would be alarmed by the amount of insecticides in and on the tobacco you buy."

As it happened, she was on to something. At the New York City offices of the P. Lorillard Tobacco Company, Plyler's note circulated around the research department. "Her letter deserves an answer," a research supervisor named C. O. Jensen said to a Lorillard vice president, "if for no other reason than to indicate our interest in the whole problem."

Tobacco companies were well aware, of course, of the hundred-plus insecticides used on tobacco, from the dusts applied to kill field insects to the sprays used on warehoused tobacco. As newer insecticides had come on the market back in the 1940s, the companies were in close contact with USDA researchers and university scientists—many funded by tobacco companies themselves—to learn how the chemicals were affecting their product. DDT was as much of a boon for tobacco growers as it was for other farmers. It killed flea beetles, which chewed little round holes in tobacco leaves, and it eliminated cutworms, which could quickly destroy a whole plant. Spraying it in warehouses kept the areas free of beetles that liked to attack tobacco while it cured. DDT also increased tobacco's burn rate and imparted a flavor that was slightly sweet, side effects that didn't seem like problems at all.

Tobacco wasn't subject to pesticide tolerances, so growers quickly got in the habit of using DDT and other pesticides with abandon. When James Delaney held his House hearings on chemicals in the food supply in the early fifties, tobacco executives had watched closely, worried that Congress might try to introduce tolerances for their crop. It didn't. A few years after the Delaney Clause went into effect, USDA scientists surveyed ten cigarette brands for pesticide residues anyway. They contained an average of 3.8 ppm DDT, an amount safely below the FDA tolerance in most foods.

In the meantime, however, a very different challenge for the industry began to emerge. A number of studies had been published in the 1950s suggesting a connection between cigarettes and cancer, and in 1957 the US Public Health Service had taken the official position that smoking caused lung cancer. The news came as cancer rates were generally on the rise and a cancer panic was settling over the nation. Tobacco companies, naturally, wanted none of the blame.

So together they had turned to the Manhattan public relations firm Hill & Knowlton, which had a well-honed set of tactics it had developed to protect the chemical industry during the Delaney hearings. With Hill & Knowlton's guidance, the tobacco industry came up with a public relations plan to protect its own products. Its flaks flooded journalists with the industry's "facts," they cultivated scientific experts critical of the existing research, and, most importantly, they began to sponsor their own medical research on smoking and health. The Tobacco Industry Research Committee (TIRC), a joint venture of the biggest tobacco firms, promised to support "the research effort into all phases of tobacco use and health." It was product protection designed to look like corporate goodwill.

By the time the surgeon general announced an investigation into smoking and health in 1962, the TIRC was already operating in high gear. Its staff tried, without success, to place industry-friendly experts on the surgeon general's panel and to discredit the scientists he chose. When *Smoking and Health* was released in 1964, it was clear that those strategies had failed. So the tobacco industry rebranded the TIRC as the Center for Tobacco Research (CTR) and gave it a revised set of mandates: to create expert witnesses willing to defend the industry in any setting, from Washington to scientific meetings to the evening news, and to publicize the idea that the science on smoking was controversial and that the findings on cancer and other forms of disease were preliminary and refutable. Over and over again from the moment of the report's release, the CTR's experts told the public that the science on smoking simply wasn't settled.

That strategy worked. Cigarette sales contracted, and tobacco tumbled in 1964 but only temporarily. By 1965, cigarette sales rose to record heights, and the industry saw its highest all-time profits.

But the surgeon general's report had also contained an almost passing mention of something that, by 1965, was starting to seem like yet another challenge for the tobacco industry. The report noted that DDT, which was used on tobacco close to the time of harvest, caused tumors in trout and rats. The report didn't say that the pesticide was the reason that cigarettes caused cancer, but it implied that it might be. By doing so, the report brought increased scientific attention to an issue that some tobacco users had been asking about for some time, since the first reports on smoking and cancer had come out in the fifties.

Now researchers at Wayland Hayes's lab in Savannah reported that DDT levels in the bodies of smokers seemed higher than those in nonsmokers. "The possible contribution of these residues to the hazard from smoking cigarettes should be explored," they wrote. Researchers at North Carolina State University surveyed DDT levels in cigarette brands anew. The average residue was now 5 ppm, higher than it had been when the USDA had done its survey five years before.

And then a new tobacco company appeared on the scene, the first new cigarette maker to join the industry in twenty-five years. In 1967 Continental Tobacco Company of South Carolina introduced the market's first proudly pesticide-free cigarette, the Venture. Continental president James Sorensen was suddenly everywhere, meeting with wealthy investors and testifying before Congress. His company had taken pesticides out of the product, he told the Senate Commerce Subcommittee, because when the company independently tested tobacco from growers, it found as much as 327 ppm endrin and 590 ppm DDT. Even cotton, he told the committee, was not permitted to have more than 7 ppm DDT.

So the industry, which had long preferred to leave federal and state agriculture departments responsible for studying the pesticides it used, began conducting its own research on the chemicals. Some of the companies started by analyzing their own products. British American Tobacco found 4.5 to 13 ppm DDT in its cigarettes. It also found that more DDT remained in cigarettes compared to other persistent pesticides. Lorillard found that DDT was present not only in its cigarettes but also in its cigarette smoke. Jensen, the Lorillard research supervisor, wrote to a friend at DDT maker Geigy for help. It wasn't much of a surprise that a pesticide as persistent as DDT survived even in cigarette smoke. The problem was, Plyler was to some extent right. The company had no idea what the pesticide meant for smokers' health.

Jensen's friend at Geigy forwarded him a large packet of everything the company knew about DDT metabolism and accumulation, along with a note. "I'd like to know how your pesticide investigation progresses," he wrote, "on a confidential basis of course."

•

Then the lawsuits began. Citizens were suing governments over DDT in New York, Michigan, and now, it appeared, Wisconsin too. None of it seemed to have anything to do with the tobacco industry, except that research departments in all of the larger companies were well aware that the pesticide was in all of their products, had been for years, and on top of that seemed to be rising.

Thus, late in 1968, executives from the nation's six biggest tobacco companies—R.J. Reynolds, Philip Morris, Brown & Williamson, American Tobacco, Liggett and Myers, and Lorillard—came together to form another industry-wide committee, the Tobacco Pesticides Committee (TPC). The committee gathered that fall in a conference room at the Statler Hilton in Greensboro, North Carolina, to discuss its shared objective, which was, as one member vaguely summarized it, to "put the tobacco industry in a better position in its relationship with the society in which it operates." The committee adjourned in agreement that this goal meant gathering as quickly as possible all of the available information on the toxicity of pesticides present in their companies' cigarettes.

The TPC members began by tracking down a list of pesticides used on tobacco globally, from a tobacco research institute in France. There were more than 106. At follow-up meetings in Louisville and Durham, they narrowed the list down to the 34 most commonly used. DDT topped it.

The committee members agreed to give themselves six months to figure out how much of each pesticide was applied to tobacco; how toxic it was when touched, eaten, and inhaled; how toxic it was over the long term; and how it was metabolized by animals and plants. It was a cooperative "interindustry" effort, with each company taking responsibility for a handful of pesticides. Brown & Williamson was assigned parathion and DDT. They all pledged to share what they found with the rest of the group, although some said, privately, that they would decide later whether or not to share what they found. Their goal was to collect everything that had been published—and everything that hadn't been—and then fund scientific experiments to see which, if any, of growers' favored chemicals were capable of causing cancer or other kinds of harm.

Over the next few months the committee members worked without stopping. The matter was urgent because if any of the pesticides did cause cancer, it would certainly harm sales if word got out. But such a finding could also give the industry a way to shield cigarettes from any further blame for the nation's cancer epidemic: they could blame the cancer that appeared to be caused by smoking on carcinogenic pesticides and then pledge not to use them on tobacco.

•

And then, very quickly, the matter became urgent for an entirely different reason. Just as the TPC had begun its work, European cigarette companies began complaining to the USDA that residues of DDT and its metabolites in US tobacco were "unacceptably high." British company Imperial Tobacco directly asked the USDA to ensure that residues in future crops were lower. West Germany, an enormous foreign market for US tobacco, said it was considering legislation to set tolerances on tobacco. If the regulations were adopted, US crops wouldn't comply.

In response, the USDA, in service to tobacco firms, turned to growers and asked them to stop using DDT. The agency told growers the reason was insect resistance: if DDT use continued at current rates, the agency scientists told growers, the pesticide would be worthless against tobacco pests in less than two years. That was true: houseflies, cockroaches, and mosquitoes were nearly all resistant to DDT by then. It just wasn't the whole truth. And at any rate, it had no effect on growers' pesticide use. DDT was among the cheapest insecticides available. They stuck with it.

So the USDA then issued a firmer threat: in 1969, while the hearings in Wisconsin were underway, the agency said it would eliminate federal subsidies for any grower that continued to use DDT. Growers shrugged. They had already sprayed DDT on that year's crop. It was too late to do any different. So the USDA made good on its threat. Anyone using DDT in 1970 would be ineligible for a subsidy from the government. The agency also announced that it was considering a proposal to make DDT's use on tobacco all but illegal.

That worked. Growers switched to other chemicals, especially the equally cheap, if less effective, DDD—the very chemical that had decimated the Clear Lake grebes.

But dropping DDT for a season was, by this time, far from enough. DDT levels in the 1970 crop still wouldn't be low enough to meet the new standards adopted by West Germany and the United Kingdom, the two largest export markets for US tobacco. "We are seriously concerned," a Georgia tobacco grower told the USDA. The plants were taking up DDT still present in soil from long-ago applications and from drift as it blew from other places it had been applied. To truly get DDT out of tobacco, he realized, it couldn't be used on other crops either. The fact was, it probably had to be eliminated altogether.

In the meantime, the TPC had collected as much information as it could find on DDT. Although the National Cancer Institute considered DDT carcinogenic, the committee members weren't convinced by the evidence—there simply wasn't enough of it, just a few studies on rodents and fish. They also found no other studies to suggest that DDT storage was consistently higher in smokers than in nonsmokers. And they found Wayland Hayes's reports attesting to the equilibrium that DDT reached within the human body reassuring. Other pesticides, such as dieldrin, seemed more hazardous. By comparison, the whole matter of DDT, said a Philip Morris executive, seemed to be "mainly a public relations issue."

Yet it was a PR issue that wouldn't go away. Even as the tobacco companies were dismissing DDT, scientists in Europe were focusing anew on its carcinogenicity. A group of scientists at the Institute of Nutrition in Budapest, Hungary, had published a multiyear study showing increasing rates of tumors and leukemia in five subsequent generations of lab mice fed DDT. The results echoed a 1947 study conducted in the US that found tumors in lab rats fed DDT. And they were troubling enough that a team of scientists from the World Health Organization traveled to Hungary to meet with the lead scientist, T. Kemény, to assess his research practices in person. The WHO scientists suggested Kemény repeat the study with more and different strains of mice. He did. The findings remained. He published the new results in English.

As news of the study spread across Europe, so did DDT panic. In Helsinki, Finland, a journalist for the country's top newspaper, *Helsingin Sanomat*, bought foreign cigarettes at local stores and had them tested for DDT. They all contained its residues. American cigarettes fared the worst—especially Marlboro, made by Philip Morris. When *Helsingin Sanomat* published the results, it noted that the "carcinogen," as the paper called DDT, wasn't just in the tobacco—it was also in the smoke.

Philip Morris had the article translated into English and then hurried to test its cigarette brands all across Europe. Its operations department collected samples from Finland, Yugoslavia, Switzerland, Italy, Germany, Holland, and Austria, shipping them to New York by airmail for private testing in an independent lab. Every last cigarette contained DDT.

Levels in the Finnish samples were high, but levels in the Italian samples were even higher. The company then tested the Swiss cigarettes for other pesticide residues and found those as well. It compared its own brands to other brands on the Swiss market. Philip Morris cigarettes had the highest residues. If the results leaked out of the lab, said a Philip Morris executive, it would destroy the Marlboro brand image.

But he did see a solution. To protect their European markets, US tobacco companies needed to get the DDT out of US tobacco. He suggested that Philip Morris ask its tobacco dealers—the companies never wanted their own fingerprints on anything—to pressure government regulators to curb the use of persistent pesticides, especially DDT. Everyone in the tobacco industry—executives in Manhattan, scientists at the USDA, growers in tobacco states, and local agricultural agents—had decided that it was time to be done with DDT.

Chapter 17

THE HEARINGS

Thomas Jukes could not believe the circus taking place in Madison, Wisconsin. As he followed the news from his home in Berkeley, California, he found it infuriating: the demonstrating students, the antics of Victor Yannacone, the young people's complete disrespect for upstanding scientific men who had devoted their entire careers to technologies that had made the world a better place to live. When he saw that Charles Wurster had accused DDT of "biocide," he could barely contain himself. This was a chemical that saved lives. Was no one going to defend it?

Jukes, a thickly bespectacled Berkeley professor, considered himself an ardent defender of wildlife and nature. He had been an active, life member of the Sierra Club for three decades. But he also cared about people. He had earned his PhD in the famous Canadian lab where insulin was discovered, putting an end to diabetes as a fatal disease. He had worked for years at Lederle Laboratories, directing the pharmaceutical company's research on vitamins, antibiotics, and anticancer drugs. The scientific contribution he was most proud of was his personal discovery that feeding antibiotics to poultry made them grow faster and stay healthy. Science was meant to serve humanity. When it didn't, starvation and disease ensued.

And that, in Jukes's view, was exactly what was wrong with *Silent Spring* and the whole antipesticide movement that followed it. Rachel Carson's book had enraged him. He found her arguments against pesticides not just antiscientific but antihumanitarian. She had never bothered to mention that DDT and the other postwar pesticides had saved more lives than just about any other medical technology ever developed. But that's exactly what they did. They were technological marvels that eliminated the insects that destroyed food crops and spread deadly disease.

The pesticide opponents now using her arguments to ban DDT were just as bad. The fact that they called themselves environmentalists seemed fitting to him. In Jukes's day, people who cared about nature called themselves conservationists. They knew that nature was precious, but they also knew that its resources were there to be managed for the benefit of mankind.

There was something else that bothered him about these new environmentalists. He had first become aware of it several years ago, when a Black woman had applied for a local Sierra Club membership and the group rejected her application. He had protested vehemently. But then he realized that despite American liberals' support for civil rights (he himself was an Englishman), environmentalists were always creating organizations exclusively for whites. The National Audubon Society was a club for the white middle class. The Sierra Club and the National Geographic Society organized tours for healthy, wealthy white people. This new Environmental Defense Fund, as he saw it, was embarking on a quest that was going to kill millions of nonwhite people all over the world. He wondered whether anyone was going to point that out.

•

In January 1969, hearings examiner Maurice Van Susteren adjourned the hearings in Madison to give DDT's manufacturers time to prepare a response. Five of the big wartime manufacturers were still making DDT: the Rothberg family company Montrose, which was now the nation's top manufacturer; followed by Diamond Shamrock Corporation; Olin-Mathieson Chemical Corporation; Lebanon Chemical Corporation; and Allied Chemical Corporation, which had just decided to drop DDT. They were all members of the National Agricultural Chemicals Association (NACA), which had learned about the hearings just a week before they began. The association had quickly hired Louis A. McLean to represent the industry; McLean was a longtime counsel for pesticide maker Velsicol Chemical Company, based nearby in Michigan.

After McLean had crumpled under Victor Yannacone's questioning, he openly told reporters that the hearings had caught the industry off guard. "Frankly," he said, "nobody knew what kind of hearing this was."

So the NACA let him go. Samuel Rotrosen—son-in-law of Montrose founder Pincus Rothberg and now president of the company—took charge. He helmed a NACA task force consisting of the five DDT manufacturers and Geigy, which participated even though it was no longer making DDT in the United States.

In meetings chaired by Rotrosen that winter, the task force came up with a strategy. They would present witnesses who would testify that banning a pesticide like DDT would decimate farmers and destroy the nation's food supply, leading to unemployment, scarcity, and hunger. They would invite scientists to testify that DDT was safe for humans, degraded quickly in agricultural fields, and was quickly excreted by wildlife. They would invite still other experts to speak to DDT's critical place in international health, for, after all, nearly 80 percent of the DDT produced in 1967 had been shipped abroad to control malaria. They would remind everyone, they agreed, that "the enemy is the pest, not the pesticide."

They also agreed to make another point, one that borrowed from the tobacco companies' strategy of calling scientific evidence into question. Back in the forties and fifties, DDT was detected in samples using something called the Schechter-Haller method, which was developed by USDA scientists at the end of the war. To determine if something contained DDT, and roughly how much, a chemist added nitro groups—nitrogen attached to two oxygen molecules—to the sample, followed by the chemicals benzene and sodium methylate. The resulting color of the solution in the test tube indicated if DDT was present, in which form, and to what extent.

In the sixties, however, chemists began using a new method of detection called vapor phase or gas chromatography. The method identified DDT in a sample by measuring the amount of time that it took for the sample, dissolved in a solvent, to move through a gas- and resin-filled column called a chromatograph. A detector at the end of the column drew a peak indicating which chemical was in the column. The method could detect much lower levels of DDT than the Schechter-Haller method could, which meant previously undetectable residues were now measurable. But DDT's peak closely resembled the peak indicating PCBs, an entirely different class of industrial chemicals that were used in a laundry list of ways for their insulating and fire-retardant properties. So the task force sought out witnesses to testify that the method could not be trusted because the chemical found in wildlife might be PCBs and not DDT.

When Monsanto executives stumbled across a news article describing this plan, they were outraged. The company wasn't on NACA's DDT task force because it no longer made DDT, but it was an influential NACA member because it manufactured so many other agricultural chemicals. It was also the nation's top manufacturer of PCBs. Its public relations staff "read the riot act" to NACA officials in Washington, where the industry group was headquartered. They could hardly believe, one of them said, that "no one in Monsanto was told of this strategy."

But the news about the hearings was also a wake-up call for the company. Its executives began to worry that other pesticides, including their own, could come under fire next. So after nixing the task force's strategy, they offered to help cover the cost of bringing witnesses to Madison to testify. It behooved every pesticide manufacturer, they agreed, to show that agricultural chemicals were a "servant of society."

When the Madison hearing resumed in April 1969, the NACA task force called its first witness: Wayland J. Hayes, now a gray-haired professor of toxicology at Vanderbilt. As he had in the Delaney hearings and the Spock trial before, Hayes again described with poise and gravitas his credentials and long history studying DDT. He carefully and methodically described the study in which he fed DDT to prisoners. The men, he said, were nothing but healthy at the end. He described a study of men who had worked at the Montrose chemical plant in California. They were healthy too, he said. Not a single man in either study showed any sign of nervous system disorders, blood disorders, lung disease, or cancer.

And then Yannacone stood up to cross-examine. The young lawyer from Long Island asked Hayes if his team of scientists had evaluated the men's liver function.

They had not, Hayes said.

Yannacone asked if they had measured levels of the men's liver enzymes.

They had not, Hayes said again.

Yannacone quickly shot off a long list of questions he knew Hayes wouldn't be able to answer affirmatively. His scientists hadn't tested DDT in infants, children, women, older people, or sick people. They hadn't tested their subjects' hormone levels. They hadn't checked to see if other enzymes were affected. They had looked no further than overt clinical signs of illness and had not bothered to update their methods with investigations into DDT's biochemical effects in the body. Yannacone waved his arms and began to shout as he questioned the toxicologist.

"Cool off!" Van Susteren interjected. "Histrionics have no place here."

Hayes remained calm and collected as he continued to explain that humans suffered no ill effects from even heavy doses of DDT. And then he was dismissed.

•

When Jukes caught the news of Hayes's testimony, he realized he could no longer stay silent. He wrote to the NACA to offer himself as a witness. He was a professor of medical physics at Berkeley, but because he had a long career behind him as a biochemist, he believed he could speak to the humanitarian benefits of pesticides, which in his view were as vital to health as antibiotics, his main area of expertise.

The task force thanked him for his offer and then politely declined. Jukes commiserated over the whole situation with his friend J. Gordon Edwards, a biology professor at San José State University.

They agreed that between the two of them, they could at least defend DDT in the press. When a local television reporter called Edwards for comment on the hearings, he ran an idea past Jukes. Should he eat some DDT on camera to show how safe it was? Jukes urged him on. Edwards did it. Jukes himself sent an editorial to the *Washington Post*. The *Post* ran it. "I am . . . sure that DDT is present in my fatty tissue," he wrote, "and I am not worried."

They were a two-man publicity team, pointing out everything they thought the public needed to know about DDT. Journalists and editors seemed receptive. Jukes wrote letters to the editor, opinion pieces, columns for top science magazines—and satirical poems that went nowhere beyond his circle of friends. (To the tune of "America the Beautiful": "Malaria! Malaria! My spleen will welcome thee! Restore the sickness grandpa knew, By banning DDT!") He sent pitches to television producers. He wrote letters to his congressmen and senators. He even wrote to Vice President Spiro Agnew and President Richard Nixon. If DDT was banned, he told them all, poor people all over the world would die. They would die primarily of malaria and starvation but also of such diseases as typhus, yellow fever, dengue, plague, and onchocerciasis, all now largely confined to the "Third World." "The issue of banning DDT," he wrote, "is unquestionably a genocidal one."

It didn't matter, Jukes said, that the present hearings concerned DDT's use in Wisconsin alone. Combined with the events in New York and Michigan, the current hearings were proof that something much bigger was likely to happen. This was all just a preview, he believed, of a much wider ban.

When he learned that the state of Washington was preparing to hold hearings too, he sensed that he had been right. He reached out to DDT's manufacturers again to offer his services as an expert witness. This time, he heard back from Max Sobelman, the onetime floor engineer at Montrose who was now a senior manager. Sobelman said that he was compiling statements in defense of DDT and that he'd be glad to accept one from Jukes. Jukes, whose typewriter was never idle, quickly sent something off. "The environmentalists, as they call themselves, have not shown sufficient involvement in basic human needs, the need for food, the need for protection against infectious diseases, to be entrusted with the responsibility of controlling pesticides," he wrote. "Have you asked yourself what a ban on DDT will do to millions of black people who can't read or write?"

His purported antiracism, as it might strain to be called at another time, was apparently also drenched in condescension.

•

Rotrosen and his task force believed that at least some of their witnesses should directly refute claims made by the scientists who testified for the EDF, perhaps especially Lucille Stickel, whose testimony had reportedly shaken everyone in the hearing room the previous winter.

In response to Stickel's testimony, NACA invited a poultry scientist who testified about experiments in which feeding 200 ppm DDT to Japanese quail had no effect on their eggshells. The study, he claimed, indicated that DDT wasn't necessarily to blame for bird declines. A wide variety of factors, he said, could affect a bird's shell thickness. "If you frighten them, you can scare the shell out of them," he said. The joke fell flat.

On cross-examination, Yannacone got the scientist to say that his expertise extended no further than poultry, which, unlike wild birds, lay eggs all year round. The two types of birds were not equivalent. Wurster was satisfied. The EDF's strategy was still working. Through cross-examination alone they were casting DDT's defenders as "narrow specialists out of touch with the rest of the scientific community."

In the meantime, Wurster did what he could outside the hearing room to advance the EDF's cause within it. He spoke with loads of journalists, wrote letters to the editor, and wrote articles about the hearings for the Audubon Society. When the *Washington Post* invited him to write an editorial, he wrote an impassioned plea for DDT's end. When the paper asked him for a photo to run alongside his words, he suggested Joe Hickey's photo of the baby eaglet. It ran on the cover of the paper's opinion section—alongside an article by

Thomas Jukes, which was accompanied by a picture of a young boy in Naples being dusted with DDT in World War II.

The Berkeley professor, whom Wurster didn't know, seemed to have it out for him personally. In the press and in the letters he sent out far and wide, Jukes called Wurster's published scientific papers fragmentary, his conclusions unfounded. Wurster's main scientific "achievement," Jukes said, was "to collect a handful of dead robins . . . [and] in a flight of fancy extrapolate these few dead members . . . into a nationwide slaughter of millions." Wurster retaliated: "Jukes' is an example of the kind of thinking that concludes all is well with pesticides as long as there is food on our tables and robins on the lawn."

But it mostly frustrated Wurster that Jukes kept bringing up human health everywhere else in the world. It was misleading, Wurster thought. In the United States, DDT was no longer used to fight disease or protect food crops. The vast majority of it—86 percent of annual production—was sprayed on cotton. The rest was sprayed on ornamentals, shade trees, and tobacco. As far as Wurster was concerned, if people in Asia and Africa wanted to use DDT to combat malaria, they could. The US, in his view, didn't need it. Malaria was gone. The case against DDT was just about the birds.

•

The Wisconsin hearings adjourned at the end of May 1969. Nearly three dozen expert witnesses had testified. The 4,500-page transcript and attached exhibits, including the baby eaglet photo, weighed 40 pounds. Van Susteren retreated to review the evidence and make his decision.

In the meantime, the EDF had unstoppably grown. The organization now had more than two hundred scientists on its Scientists Advisory Committee, a group at the ready to testify and provide scientific statements as needed. Communities all across the US were now calling on the organization for help in battling their own local environmental problems: air pollution in Missoula, an oil leak in Santa Barbara, underground radiation in Denver. But for all of that growth and demand, the organization was still short on cash.

Its only members were its board of trustees, and the board worried that its current model—Wurster and Yannacone doing the lion's share of work in a sequence of never-ending, last-minute, late-night frenzies—just wouldn't hold.

In the summer of 1969, the EDF's board hired an executive director and then attempted to shift Yannacone to the position of legal counsel. It was a demotion for Yannacone, who bristled. Lorrie Otto recalled times early in the Madison hearings when Yannacone was hurt by things that EDF members had said about him. She knew that Yannacone had accused some in the EDF of slandering him and that Wurster and others in the EDF found Yannacone greedy and demanding. Although Yannacone and Wurster respected each other, she could see the tensions between them coming to a boil.

A short time later, the board removed Yannacone and cut all ties. The board offered several explanations: Yannacone's "lack of respect" for them, his brash courtroom style, his high monthly retainer, and his insistence on bringing a suit asking for $30 billion in damages from DDT manufacturers as a class action on behalf of all US citizens. Yannacone would not relinquish the suit, and the board wasn't fond of it. Yannacone left quietly, certain he had been pushed out for racial reasons: because he was an Italian who cursed in Yiddish, he said, and because he had worked for an organization that defended Blacks.

Wurster, still in place, decided to push the EDF to stay on top of the DDT issue no matter what happened in Wisconsin. He also had an idea about how the EDF should tackle the pesticide in future actions. If its defenders—like Jukes—were going to bring human health into the picture, then the EDF should do the same. The timing for this shift in strategy, he said, was right. The National Cancer Institute had just released study results showing that DDT caused liver tumors in mice, and the findings were stronger than those from earlier studies linking DDT to cancer. This study was carefully conducted, the tumors it found were malignant, and the design of the study revealed that DDT caused as many tumors as the weed-killer aminotriazole, a known carcinogen. Then the FDA announced a ban on artificial sweeteners called cyclamates in soft drinks, based on evidence that the chemicals caused cancer in lab mice. Under the Delaney Clause, the FDA ruled, cyclamates had no place in the food supply. A week later, seventeen members of Congress told President Richard Nixon that by the same reasoning, it was time to ban DDT.

Wurster grasped that the question of carcinogenicity, properly handled, could alter DDT's fate at a national level. He shared an idea with his EDF colleagues. It was time, he said, to start arguing that DDT was like cigarettes, carcinogenic, even if the extent of its carcinogenicity might not be fully understood for decades to come.

His colleagues agreed. That fall the EDF, in partnership with the Sierra Club, the Audubon Society, and a Michigan environmental organization, petitioned the secretaries of the USDA and the Department of Health, Education and Welfare to cancel DDT's registration and set its tolerance in food to zero. Then, with a California group that provided legal counsel to the United Farm Workers, they filed a petition against the Health, Education and Welfare Department on behalf of pregnant women, also seeking a tolerance of zero in foods. Because DDT caused cancer in mice, they argued, it had no place in breast milk or food.

To their surprise, the agencies responded quickly. In November 1969, Agriculture secretary Clifford Hardin announced that within ninety days, DDT would no longer be permitted in and around homes or on marshes and wetlands, shade trees, or—curiously—tobacco. Health, Education and Welfare secretary Robert Finch announced that same month that over the next two years, all nonessential DDT uses would be eliminated; Finch made the announcement in response to findings of an independent pesticide commission he had created, chaired by food scientist and longtime University of California, Davis Chancellor Emil Mrak. But Finch, following the commission's finding, rejected the zero-tolerance request as impractical. New technology made ever-smaller traces of pesticides detectable, the commission had argued, which meant it might never be possible to achieve residues of zero.

The announcements marked the first time the federal government took steps to cancel a pesticide's uses because it was inherently dangerous, not just mislabeled or misused. Yet the EDF believed the federal agencies should go further. At the very end of 1969, the organization petitioned the US court of appeals in Washington, DC, to direct the secretaries to take "more effective" action.

As the EDF waited for a reply, it pushed forward with a plan to open up its membership to the public. Early in 1970, the organization ran a big, pricey ad in the *New York Times*. It depicted a mother nursing an infant. "Is Mother's Milk Fit For Human Consumption?" it read. "If it were on the market it could be confiscated by the Food and Drug Administration. Why? Too Much DDT."

Inspiration for the ad came from Swedish scientist Goran Lofroth's testimony. It represented a shift for the organization in more ways than one. It opened the EDF up to the public, and it sent the public a very specific message about the organization's mission: the EDF wasn't just worried about birds, fish, and wildlife. It was an organization that cared about human health above all.

Later that spring, the organization heard back from Van Susteren in Madison. DDT, its breakdown products, and other chemicals similar in structure, he found, were environmental pollutants by the definitions laid out in Wisconsin statutes. They harmed birds, fish, and other forms of animal life. And in the environment, he determined, there was no concentration or amount that could be considered safe. The EDF rejoiced.

In the meantime, Yannacone had filed his class-action suit. He gave a talk at a massive environmental teach-in at the University of Michigan in Ann Arbor. It was a weeklong event drawing fifty thousand people, a prelude to the first Earth Day, which Wisconsin senator Gaylord Nelson had pledged to organize that spring.

Nelson spoke to the crowds in Ann Arbor, as did Ralph Nader, United Auto Workers president Walter Reuther, Hollywood stars, and corporate executives. But Yannacone's rousing speech set the theme for the week. Corporations had advocates, he said. Industries had advocates. Government, political organizations, and the industrial-military power structure all had advocates. People needed advocates, he said, and they needed the law. Industry would ignore protests. Government would ignore protests. "But no one in industry or government ignores that scrap of legal cap that begins, You are hereby summoned to answer the allegations of the complaint annexed hereto within twenty days or judgement will be taken against you for relief demanded," he said. "Don't just sit there and bitch. Sue somebody!"

Chapter 18

DESTRUCTION

Jack Hartsfield was what people called a newsman. He had covered crime in San Bernardino and the Moon landing from Cape Canaveral. He had also worked in public affairs for the air force, so he knew whom to call when he had to report a story about the armed forces—which, these days, was often. He had recently moved back to his home state of Alabama to work as the science editor for the *Huntsville Times*, and a good number of his stories came from the US Army's Redstone Arsenal, home to NASA's Marshall Space Flight Center.

Increasingly, though, many of the biggest science stories of the day weren't space stories but pollution stories: fish kills, river fires, urban smog, lakes and creeks too clogged with garbage to drink from, swim in, or even just look at. Earth Day had proved that people—especially young people—wanted the pollution to stop; the event had exploded into countless antilitter and antipollution rallies and teach-ins all across the country. So when Hartsfield heard that the Redstone Arsenal had a pollution problem, he started asking around. It was just the kind of story he needed to follow.

He learned that new federal water-quality standards authorized by Congress in the sixties were causing problems for the army site. The arsenal was home to the nation's only DDT plant on federal land, a plant leased and operated by Olin Chemical Corporation.

And the plant was contaminating the nearby Wheeler Wildlife Refuge with DDT.

Wheeler was one of the first wildlife refuges in the nation. Tens of thousands of migrating ducks and geese overwintered in its waters, which included the Tennessee River and tributaries that flowed across the Redstone Arsenal. In winter, Wheeler's bird population could reach 150,000 birds, some from as far as Ontario. Birders spotted blue-winged teals and other dabbling ducks, Canada and snow geese, American coots, sandhill and whooping cranes, white pelicans, and even the occasional bald eagle. Until just recently, some of Wheeler's tributaries, including the Huntsville Spring Branch, one of the tributaries on the arsenal, had also provided drinking water for some of the small towns outside of Huntsville. And now all of that water was heavily contaminated with DDT.

But the story that Hartsfield wrote up for the front page of the *Huntsville Times* in April 1970 wasn't about the contamination—that, he learned, was old news. The paper hadn't covered it, but the army had known since the 1940s that there was DDT in the tributaries on the arsenal: the Huntsville Spring Branch and Indian Creek. The army had also been pressuring Olin to stop the contamination for years, and the company had tried to solve the problem with catchment ponds, filtration systems, and landfills, although none of it had worked. When Hartsfield caught up with the story in the spring of 1970, while states up north were holding high-profile hearings on DDT, he learned that Olin and the federal government were in a protracted battle over the plant. Olin was requesting permission to continue its manufacturing under older water-pollution standards. The Federal Water Quality Administration, the Interior Department division responsible for enforcing the newer standards, knew that Olin would never be able to meet them. But if the administration gave Olin a pass, then the federal government wouldn't be able to enforce its new standards anywhere.

Hartsfield's story made for a banner headline in the Sunday paper. Drinking water at the arsenal contained DDT, he reported. DDT levels in the streams on the arsenal were nearly two hundred times higher than the new federal water-pollution standards allowed. Some of the drainage ditches outside the Olin plant contained as much as 74 percent DDT. Wildlife specialists he interviewed said the pesticide was probably behind local declines in bird populations and the disappearance of local mussel fisheries. It was also likely why, in downstream towns like Triana, it had become nearly impossible to spot turtles, frogs, and other forms of aquatic life. What this all meant for people, Hartsfield wrote, was still unclear, although medical experts did know that the chemical accumulated in their fat.

Triana mayor Clyde Foster didn't catch Hartsfield's story, which ran eight years before his eruption in 1978. Foster, always a busy man, hadn't had a moment to spare in all that time. The Army Ballistic Missile Agency, where he was hired back in 1957, had become the Marshall Space Flight Center when NASA was created, and the center had spent the last decade developing the Saturn rockets that had just helped launch the Apollo 11 and its astronauts to the Moon. In his job at the center, Foster had been writing the mathematical models and computer programs used by NASA engineers behind the scenes. He had also convinced his supervisors to lend him out to Alabama A&M, his alma mater, to start a bachelor's degree in computer science there. On his own dime, he was traveling to other Black universities to recruit scientists and engineers to come to Marshall, too. In the meantime, as mayor, he had brought Triana running water by deeding part of his own property for the pumphouse, part of his never-ending work to bring the town into the present.

The disparity between Marshall and his town rankled all the while. When Triana had celebrated Apollo 11's return in the summer of 1969, he hadn't been able to look away from the bales of cotton stacked outside the county courthouse.

"Bales of cotton!" he said. "And here we'd placed a man on the Moon."

•

So Foster didn't catch Hartsfield's story, but the *New York Times* did. And that was where Charles Wurster read about it early in May

1970, just after the EDF had opened up its membership with its big ad about DDT in mothers' milk.

The *Times* article caught Wurster's attention because it pointed out that a DDT plant on federal land was contaminating protected federal land. A bird sanctuary, no less. And there was more: the refuge supplied the drinking water for the Alabama city of Decatur. Fishing was popular in the river. Wurster, on a constant lookout for ways to keep the EDF's DDT battles front and center, thought that the organization had to act.

He got in touch with Edward Lee Rogers, the EDF's new attorney. Wurster pulled together the same scientific evidence he had assembled for the Wisconsin hearings; he knew it was airtight. Once again, the organization prepared to sue—but not Olin. On June 5, 1970, the EDF, working with the National Wildlife Federation and the National Audubon Society, filed suit against the federal government for violating its own water-quality standards by allowing discharges of DDT-contaminated wastewater into the wildlife refuge.

Wurster was stunned by the response—from Olin. Three days later, on June 8, Olin president Gordon Grand made a public announcement: the company would close its Redstone Arsenal plant at the end of the month. It was Olin policy, Grand said, to comply with all present environmental standards.

Grand didn't mention that state officials were already after the company for polluting two rivers in southwestern Alabama, where its McIntosh plant produced chlorine for its DDT. He didn't mention that the federal Justice Department was considering whether to authorize suits against the company for discharging mercury into Georgia's Savannah River and the Niagara River in New York. He also didn't mention that Olin was already facing two lawsuits, one of them a $50 million class-action suit, over allegations that its McIntosh plant had discharged mercury into Alabama's Tombigbee and Tensaw Rivers.

Olin manufactured plenty of far less controversial things: other pesticides, fertilizers, aspirin, chlorine tablets for backyard swimming pools, cupro nickel for quarters, and gunpowder, among other products. "Anybody can make money in this kind of economy," Grand had said after ascending to Olin's presidency a year before. The real question, he said, was "Do we have the kind of operation that can turn a profit when things aren't so good?" With bad publicity only getting worse, it was easy to conclude that Olin no longer needed to make DDT.

But the manager of Olin's Redstone plant, B. H. Wilcoxon, saw things differently. The plant, he said, had "no records" showing that its operations had ever polluted the refuge or the river. And now seventy-five locals were out of jobs. Forty-four of them signed an open letter to the *Huntsville Times*, complaining that the "rumors, half-truths and vicious lies" in Hartsfield's article had cost them their life's work.

Up in New York, Wurster and his EDF colleagues celebrated what looked to them like a record-fast victory. With Olin's Redstone plant shuttered, there were now just three big DDT manufacturers left in the United States: Diamond Shamrock, Lebanon, and Montrose. This particular suit felt like a real turning point. For the first time, Wurster said, "we had won by winning."

•

Grand might have imagined that Olin would be out of the DDT business sometime in the early seventies—just not perhaps quite as early as 1970. The company terminated its lease with the army. An army wrecking crew knocked down the plant walls and tore everything to the ground. By 1971, nothing remained but rebar and rubble.

That, it turned out, was the easy part. The DDT on-site still had to be dealt with. No one knew how to speed up DDT's breakdown, but scientists at the Federal Water Quality Administration recommended combining it with lime and ferrous sulfate, a decades-old water-treatment-plant protocol that seemed to break the chemical down in lab experiments. A cleanup crew set to work on the settling ponds and drainage ditches that circled the defunct plant, still full of DDT. The crews dumped in lime, stirred in ferrous sulfate, and covered it all with heavy clay and topsoil. Then they planted the burial sites thick with grasses to prevent runoff and erosion.

Olin still had plenty of DDT inventory to unload: fourteen different products containing the pesticide, most of which the company sold to cotton farmers. The rest it shipped overseas, mostly to India, Pakistan, and Brazil, which were still using the pesticide on crops and to fight malaria. With that, the company's last stocks rapidly dwindled. But if any of its engineers or technical experts knew about the Clear Lake study, they might have warned that the chemical was not going to be so easy to get rid of.

•

That summer, President Richard Nixon signed an executive order creating a new arm of the federal government, the Environmental Protection Agency. Nixon, hoping to broaden middle-class support for his administration, had been angling to be seen as an advocate for the environment. Not long before, he had supported Congress's approval of the National Environmental Policy Act, which required all federal agencies to consider the environmental effects of their actions. But the creation of this new agency stemmed from a recommendation of his own Task Force on Resources and the Environment, which his administration had convened shortly after he was inaugurated in 1969. The task force had found that environmental activities and policies were scattered across nearly every government agency, with no single unit responsible for their overall regulation or impact. The federal government, the task force concluded, needed an environmental "focal point."

With Nixon's executive order, the EPA, an amalgamation of offices from other agencies, including the USDA, the FDA, and the Department of the Interior, became that focal point. That December, Nixon appointed its first administrator, Republican William Ruckelshaus, an Indiana lawyer and congressional representative who had spearheaded water- and air-pollution controls in his home state. He was stepping into a job and an agency with no precedent in the history of the federal government. As he told a House committee, he would approach the position with complete independence, leading the agency unencumbered by the obligations to agriculture or trade that other agencies bore. The EPA, said Ruckelshaus, had "only the critical obligation to protect and enhance the environment."

Pressure was mounting to do so by tackling the pesticides problem. A recent investigation by the General Accounting Office had found that the USDA's Pesticide Regulation Division had not responded to violations of the federal insecticide law FIFRA in over a decade. The House Committee on Agriculture was fielding a stream of legislative proposals to ban certain pesticides or overhaul regulation. Absent federal action, some states and cities were passing their own bans, with DDT the popular target. Ruckelshaus—known to friends and colleagues as Bill—now ran the singular agency that, for the first time, brought federal pesticide policies from the USDA, the FDA, and the Department of the Interior under one roof, in the new EPA Office of Pesticides Programs. The move was a long time coming, even if it was, as one political scientist put it, "a shotgun marriage of longtime bitter rivals."

•

A few months later, in early 1971, the fishery chief of Alabama's Game and Fish Division, Archie Hooper, got word of a fish kill near the Redstone Arsenal. The division was responding to reports of fish kills on a regular basis now. A kill on the Tombigbee River down in McIntosh in 1968 had left 50,000 fish dead. A few months after that, another 50,000 turned up. In 1969 his team counted more than 750,000 carcasses in a kill near Birmingham. The culprits in all three cases were the pesticides diazinon and malathion. But over in Tombigbee, Hooper's scientists had also found largemouth bass with 13.33 ppm DDT. It was easily one of the highest DDT levels he had heard of in the Southeast.

When Hooper's team went up to investigate the kill near the arsenal, they already suspected pesticides. Even so, they weren't prepared for the DDT levels they found in the fish there. At 313 ppm, the levels were nearly twenty-five times higher than those in the Tombigbee bass. This *was* the highest residue Hooper had ever heard of.

The finding struck him as a bit strange, too, because the arsenal's DDT plant hadn't been operating for months. Yet there were so few dead fish compared to the kills elsewhere. So he documented the kill and moved on.

Chapter 19

THE BAN

While Archie Hooper was puzzling over the fish kill near the arsenal, in January 1971 the US court of appeals ordered the brand-new EPA to immediately cancel all of DDT's uses. EPA head Bill Ruckelshaus complied. But federal law granted EPA the authority to suspend all of DDT's uses only if it was proven to pose an imminent hazard to people or the environment. After a sixty-day review period, he concluded that although several of DDT's uses had been canceled, by law the chemical could still be registered as a pesticide—and remain on the market—because it had several uses whose benefits outweighed its risks.

Both of Ruckelshaus's actions—his compliance with the appeals court and his later reversal—put the new agency squarely at the center of the boiling DDT issue. DDT manufacturers and formulators had requested a review of the initial cancellation order, which they were entitled to under the Federal Insecticide, Fungicide, and Rodenticide Act (FIFRA) and still wanted. And the EDF, emboldened by one DDT victory after another, responded to Ruckelshaus's reversal by suing the EPA.

Ruckelshaus's agency responded with an announcement: the EPA would address all of the relevant actions—the EDF's petitions, its lawsuit, and the industry's hearing request—by holding a single, consolidated DDT hearing. The main question of the hearing would be whether DDT's continued uses were still allowable under FIFRA, which held that an "economic poison" could not be registered for use if it injured "man and other vertebrate animals, vegetation, and useful invertebrate animals." If it caused injury or was "misbranded"—misleadingly labeled—its registration had to be canceled under the law.

The Consolidated DDT Hearings began in August 1971 in Alexandria, Virginia, overseen by a hearing examiner named Edmund M. Sweeney, a lawyer dispatched from the US Civil Service Commission because the EPA had no examiner of its own. The hearings echoed the proceedings in Madison but on a grander scale. The manufacturers invited Wayland Hayes in to speak about DDT's safety. They called in US surgeon general Jesse Steinfeld, appointed by Nixon in 1969; agricultural scientist Norman Borlaug, winner of the 1970 Nobel Peace Prize; and Thomas Jukes. Steinfeld and Borlaug testified that DDT was critical for public health abroad and that a US ban would taint the chemical's worldwide reputation, forcing foreign governments to abandon it too. Jukes planned to make the same argument but had little opportunity. He flew from California, took the stand, and began to answer questions. But after determining that Jukes was discussing only DDT research conducted by others, not himself, Sweeney thanked him for his time and politely dismissed him.

For neutral witnesses, the EPA brought in two scientists from the World Health Organization's International Agency for Research on Cancer. Its experts made noncommittal statements. The chemical caused cancer in mice, they said, but while this indicated possible carcinogenicity in people, it didn't prove it. Scientists from the National Institutes of Health then described experiments in which rats and beagles fed DDT showed no sign of tumors. The evidence on cancer was coming across as far from definitive.

Charles Wurster had anticipated this. In his own affidavit for the hearings, he expanded on his cigarette analogy. It had taken scientists decades to show the connection between smoking and cancer, he pointed out, and they were only able to do so by comparing smokers with nonsmokers. Because DDT was now in every American's body, that would never be possible. But the surgeon general's 1964 report on smoking had provided new ways of thinking about cancer causation. Only now did doctors and scientists know that they could observe dozens of smokers for years and not observe a single case of cancer among them—and this didn't mean cigarettes were not carcinogenic. Cancer agents could set off an irreversible cascade of biological events at any time. Even a small, brief, one-time exposure to a carcinogen could cause cancer. That meant that it was impossible to establish a safe, "threshold" dose for any carcinogen.

"The frequency or incidence of cancer induction becomes zero only at zero concentration of the chemical," Wurster wrote. He did not mention that carcinogenesis was far from his area of expertise.

He didn't have to. The EDF lined up witnesses who testified to DDT's carcinogenicity. They had already assembled the relevant evidence for their petition on behalf of the women they were representing with the UFW in California: by now, it included the National Cancer Institute and Hungarian studies, and a study by researchers at the University of Miami who had measured DDT levels in cadavers. The levels were twice as high in people who had died of cancer compared with people who had suffered sudden or accidental deaths.

That study, the counsel for the NACA said, proved nothing because it didn't indicate that DDT caused the cancer in those who had died of the disease. In a closing statement, the NACA lawyer focused on the rodent studies instead—but only to dismiss them. "We find it inconceivable," he said, "that Respondent and Intervenor EDF have, in effect, placed 72 mice ahead of a quarter of a century of DDT's usage unsurpassed in the history of man in terms of safety as a chemical pesticide."

A good part of the hearings grappled with the question of cancer, but a far greater part was devoted to DDT's wildlife effects. Scientists testified to the reproductive declines among brown pelicans on California's Anacapa Island and peregrine falcons in Alaska, and to eggshell thinning in thirty-nine species of fish-eating birds in North America. Sweeney pressed the wildlife witnesses hard during cross-examination. A reporter covering the hearings for *Science* wrote that Sweeney, who had previously worked for the railroads and the Bureau of Mines, was favoring witnesses for industry and insulting scientific experts. Sweeney, offended, had the reporter's article entered into the record.

The hearings went on for seven long months. It wasn't a quiet time for the country. The war in Vietnam dragged on, riots broke out in Attica, and union workers went on strike at docks and coal mines across the country. Lawmakers in Washington battled over busing schoolchildren to desegregate schools. Nixon traveled to China to reopen diplomatic ties to the Communist nation. After the last DDT witness had been called, the consolidated hearings came to a close. The transcript ran to more than nine thousand pages. On April 25, 1972, Sweeney made his recommendation: he was convinced of DDT's benefits for public health and its importance to agriculture, and he was not convinced of any resulting harm to humans. There was no need to further cancel DDT's registration.

But the final word belonged to Ruckelshaus, and Sweeney's recommendation was far from his only consideration. His agency, the EPA, was working to revise federal pesticide law so that it served not only manufacturers and users but also the environment. The federal government had also commissioned its own reviews of pesticide safety in the meantime, one at the behest of the Health, Education and Welfare secretary and one at the behest of Montrose. Ruckelshaus took it all into account. Nearly two months after Sweeney made his recommendation, on June 14, 1972, Ruckelshaus issued an executive order that overruled it.

DDT's persistence and biomagnification in the food chain made it an "uncontrollable" chemical, Ruckelshaus's order stated. The possibility that it might cause cancer and its "unknown and possibly forever undeterminable long-range effects" were grave concerns. The order affirmed DDT's cancellation. It granted an exception for controlling vector-borne disease and body lice. But any DDT manufactured and shipped over state lines in the US had to bear a label stating, "THIS SUBSTANCE IS HARMFUL TO THE ENVIRONMENT." Manufacturers who wanted to ship it abroad, by contrast, could do so as they wished.

Environmentalists celebrated. The EDF claimed the decision as a victory.

No one asked whether the tobacco industry had anything to do with it.

•

Somewhere in the midst of all that, a rumor began. Some said it started with a talk that Victor Yannacone gave at a luncheon, not long after he left the EDF. Jukes heard the story from his friend Edwards at San José State. Edwards said Yannacone told it to him directly. Yannacone, according to Edwards, insisted that Wurster had said some truly pernicious things. Someone then passed the account on to Representative John Rarick, a Democrat from the cotton-producing state of Louisiana.

Rarick was a member of the House Committee on Agriculture, and in committee hearings on a proposed FIFRA amendment held in the spring of 1971, he had the opportunity to question a new member of the EDF's legal team, general counsel Edward Lee Rogers. Rarick interrupted his own line of questioning on the EDF's operations and funding to ask Rogers a few questions about Wurster. "Would this be the same Dr. Wurster," asked Rarick, who said, "'People are the cause of all the problems. We have too many of them. We need to get rid of some of them and this is as good a way as any'?" He went on. Would this be the same Dr. Wurster who said, "It doesn't really make a lot of difference," if pesticides like organophosphates replace DDT "'because the organo phosphate acts locally and only kill[s] farm workers and most of them are Mexicans and Negroes'?"

Rogers bristled. "This is a very serious matter that you are bringing up here," he told Rarick. He insisted that it was "notorious hearsay." He said he seriously doubted Wurster had ever said such things.

When Rogers stepped down, he quickly called Wurster and told him to write to the committee in his own defense. "Dear Congressman," wrote Wurster, "I wish to deny all of the statements."

It was too late. The alleged remarks were already in the *Congressional Record*. Later—decades later—people who believed that DDT had been wrongly banned would find them there and haul them out.

Chapter 20

TRIANA

For five years, Olin's DDT waste sat buried beneath the clay on the Redstone Arsenal while the grasses grew and the remaining rubble from the plant slowly turned to dust. Then, in the summer of 1977, engineers from the Army Environmental Hygiene Agency, which oversaw the army's public health needs, ran a routine water-quality survey at the arsenal. The agency's surveys had always detected DDT in the arsenal's water supply, but now the level was higher than usual, which made little sense. DDT from the old plant had been buried for years. With the lime and ferrous sulfate treatment, it should have broken down. The agency's engineers wondered if it was coming from somewhere else, and they wondered where else it was turning up.

The engineers collected water and sediment from the Huntsville Spring Branch just below the old plant site. These samples also contained DDT. The engineers collected catfish, bass, and bream from the branch and from Indian Creek next, following the water all the way down to where the creek met the Tennessee River at Triana, about six miles southwest of the arsenal site. Some of the fish down there had as much as 412 ppm DDT, even higher than the record-setting level Archie Hooper had documented six years before.

The army engineers shared the findings with the Tennessee Valley Authority, which managed the big hydroelectric dam on the Tennessee River. The TVA's scientists had also found high levels of DDT in the area and its wildlife. They summed up the findings in an agency report they titled "Where the Water Isn't Clean Anymore." When scientists at the federal Fish and Wildlife Service in Patuxent, Maryland, saw the TVA report, they went down to Alabama to attempt to learn what the findings meant for the Wheeler Wildlife Refuge. They collected fist-sized chunks of crystalline material they found on the banks of the creeks. They collected mallard ducks, crows, swamp rabbits, and earthworms. The crystals were 21 percent DDT. The birds had DDT levels known to cause eggshell thinning. The rabbits had levels higher than the FDA tolerances allowed for the food supply. And the worms had 224 ppm DDT, a level high enough to kill birds outright.

The DDT buried at the site years before hadn't degraded after all. Instead, heavy rains had leached it from the site and carried it into the tributaries, where it was now working its way into the food chain anew.

The fish and wildlife scientists informed the army's hygiene agency, which asked Redstone officials to notify state officials. The public needed to know that fish from the creeks and the river weren't fit to eat. Then the agency called a press conference at the arsenal. DDT levels in local fish and waters far exceeded federal limits, they announced. The streams had clearly been contaminated since the 1940s. And although the arsenal's DDT plant was history, DDT levels in the environment were climbing.

The arsenal rag the *Redstone Rocket* covered the press conference. A couple of newspapers in the region gradually picked up the story too. And that's how Triana mayor Clyde Foster first learned that people in his town were eating way more DDT than the government thought was good for anyone.

•

When Foster started making calls, he quickly learned that the army had known about the pollution for years, possibly decades. He learned that the Federal Water Quality Administration, now operating under a new name as part of the EPA, had been warning about the pollution for at least eight years. And he learned that the EPA found the DDT levels high enough to warrant closing a two-mile stretch of the Tennessee River. It was the stretch where he and just about everyone else he knew in Triana loved to fish.

Foster also learned that the state of Alabama disagreed with the EPA. They knew about the pollution, too, and had refused to close the river. State health officer Ira Myers said there was no need to close it because such an action wouldn't be enforceable and because the pesticide was harmless. But Foster knew it was a federally banned substance. How harmless could it be?

No one he called at the state, the EPA, or the TVA could provide a definitive answer. But when he reached an office at the CDC, he learned that the agency could possibly help figure out the answer if he filed an official request for the agency to get involved in state affairs. As mayor, he had the authority to do just that. So he did.

•

Foster's request eventually landed on Kathleen Kreiss's desk at the CDC in Atlanta, the signatures of twenty officials well above her in rank already on it. The mayor's request was described on an agency form known as an Epi-1, which authorized the CDC, a federal agency, to get involved in a health matter in one of the states. The form identified Kreiss as the federal officer who would travel to Triana, Alabama, to collect samples of human serum and have them analyzed for DDT.

Kreiss was new to the CDC, part of a corps of investigators known as the Epidemic Intelligence Service (EIS). The service had been established back in the fifties by a CDC scientist named Alexander Langmuir to sniff out instances of biowarfare during the Korean War. When Kreiss entered Harvard Medical School in the seventies, EIS officers were tackling health problems around the world—famine in Biafra, smallpox in India, the emergence of a new disease, Ebola, in Zaire—and Langmuir was teaching an elective on the EIS at the school. Kreiss enrolled. She got hooked, finished her residency, and moved to Atlanta to join the service. That had been just five months before.

Now, looking at the form, Kreiss had to admit that she knew little about DDT. She'd been an antiwar protester in college, and she knew that there had been controversy over DDT when she'd been in medical school, but she'd been too busy to follow it. She had only a vague memory of the news on its hearings and subsequent ban. Now here she was, six years later, tasked with figuring out whether it made people sick. She started looking into the literature and soon saw that a CDC toxicologist before her time, a man named Wayland J. Hayes, had fed DDT to prisoners and then studied them. He was a professor at Vanderbilt now, nearing retirement. She read his studies and called him.

Hayes took her call, but Kreiss quickly gathered that he wasn't going to be much help. It was ridiculous for CDC to investigate the situation, he told her. DDT was absolutely harmless. You could put it on your breakfast cereal, he said.

But the Epi-1 was signed, and investigating was her job. By now, the EPA and TVA had also asked the CDC to determine whether there were any health problems related to DDT in Triana. She thanked Hayes for his time and set off.

•

Hayes couldn't have predicted the DDT levels Kreiss found in the first twelve Triana residents she tested. No one could have.

There was no DDT in Triana itself; it was only in the creek water and the fish. And Kreiss, with Foster's help, had intentionally selected twelve people who had never worked in agriculture or at Olin's plant. Yet the people's DDT levels were record setting, higher than levels of people who had worked for years in manufacturing plants elsewhere. One man, good-natured Felix Wynn, eighty-three, who wore spectacles and favored three-piece suits when he wasn't fishing, had a DDT level four times higher than the highest level ever recorded anywhere in the world.

Back at her CDC office in Atlanta, Kreiss talked over the findings with her colleagues and a mentor, Renate Kimbrough, a German physician turned toxicologist who was as sharp as a tack, full of strong opinions, and one of the few women in the field. Everyone had DDT in their bodies, Kimbrough told her, but this was indeed exceptional.

Poring through the literature, Kreiss realized something else. Hayes had studied people who had eaten DDT for no more than four years. Other scientists had studied factory workers exposed for longer periods, but the workers had either handled or breathed DDT—none of them ate it. The people in Triana, however, had been eating DDT-laden fish for twenty-three years—nearly a generation. And the blood tests had shown that DDT levels were highest in those who ate the most fish. No one knew what DDT could do to a person who ate that much for that long.

That was one reason the town seemed worthy of further study, but there was another reason, too. No community with so many women and children with such high levels of DDT had ever been studied. And yet evidence from nature, especially birds, showed that DDT's metabolites—the exact form of DDT that was so high in the people of Triana—could act like the female hormone estrogen. Triana offered a unique opportunity to figure out whether DDT could affect fertility or reproductive health in people.

Kreiss proposed a follow-up investigation of the people in Triana. She calculated the cost of a second study, just over $120,000, wrote up her justifications for it, and summarized it all on a new form, an Epi-2. When she had collected all the signatures she needed for approval, she invited the entire town of Triana and those in its rural surrounds to volunteer samples of their blood, urine, and breast milk. She was stunned by the response. More than half of the thousand people she invited—518 people—signed up.

Mayor Foster had everything to do with it. He had held town meetings and gone door to door, convincing families to participate one by one. Kreiss herself got caught up in his persuasiveness. She assembled a big team to work on the study. He set up an ethics committee of his residents to review the study design. He asked that the data from the study be shared with the University of Alabama in Huntsville and the director of a Triana health facility—even though one didn't exist yet. No matter: she wrote the nonexistent facility into her request for approval. Soon, even Foster's second-oldest child, Edith, a medical student who had worked as a phlebotomist at Huntsville Hospital, was on Kreiss's study team. Edith put on her scrubs and prepared to draw blood from her family members, neighbors, friends, and townspeople she had never met.

The study took place in May 1979 in downtown Triana's Municipal Building. Kreiss and the other two physicians she brought with her sat down with each resident and went through a long list of questions. Did they smoke? Did they drink? How much? How many times had they been pregnant? How many children did they have? The physicians took blood-pressure readings and urine samples, and then asked each person to fast and return the next morning for a blood draw. The team needed two tablespoons of blood from each adult, but less from children. Ultimately, a third of the samples they collected were from children, the youngest just a few weeks old. Their parents desperately wanted them checked. Another forty or so samples came from commercial fishermen and their families, many of whom lived miles from Triana. Because they fished for a living, they were just as intimately connected to the Tennessee River.

A photographer for the *Huntsville Times* came down to Triana that week to take pictures. He captured Edith Foster filling vials with the blood of her friends and neighbors. For Edith, it was strange to think of their tributary of the Tennessee, Indian Creek, as polluted. They ate fish from the river several times a week. A great many people in Triana did. Her mother had grown up in Triana using the creek for bathing and cleaning. They called the days they went down to wash "River Day." The water wasn't fit for drinking, but it was fit enough for everything else—or so they had thought.

Once an hour, an assistant on Kreiss's team ran the blood samples up to the arsenal, where a commander at the army hospital there had provided them lab space for the week. Lab assistants centrifuged the blood samples and pipetted them into vials. At the end of the week, they packed the hundreds of samples on dry ice and sent them back to Atlanta. There, back at the CDC, Kimbrough was waiting to help Kreiss figure out exactly what was in them.

•

The TVA, which had sprayed northeastern Alabama generously with DDT to kill mosquitoes back in the forties, offered its own aid to Triana. The agency gave the town two boats with fifty-five-horsepower inboard engines and a cache of fishing gear. They paid a crew to take the boats up the Tennessee to cleaner waters each day so that everyone in Triana could line up in the afternoon, under the shade of a tree near the fire station, and pick up fish for their household for free. The TVA also gave the town seeds, fertilizer, and gardening tools to help make up for the loss of fish from their diet. It all reflected the agency's "major new goal," recently announced, to connect with people as it worked to reduce dependence on oil, to care for the environment, and to spur economic development.

People were grateful for the gestures, but the gestures did little to stem the anger, fear, and confusion that welled up inside them. Ever since the results on the first twelve Triana residents had come back, they all knew they had DDT in their bodies. And even though those twelve hadn't worked in the plant, some Triana residents *had*. Also, many of them had worked in cotton production; it was the town's main source of employment. What if their levels were as high as Mr. Wynn's—or higher? The worry grew as they heard, over and over again, that no one knew what DDT could do to you—but that their own bodies might settle the question.

Then the commercial fishermen from up near the Whitesburg Bridge, who were mostly white, began complaining about Triana's special treatment. Their own losses were just as bad, they said, if not worse. The fishermen started talking about filing suit against the army or Olin. All the while, reporters from Montgomery, Birmingham, Anniston, and even out-of-state papers and magazines came through. They asked questions, looked at the abandoned river, and gawked at the small town's predicament.

•

Later that summer, Mayor Foster got a call from Kreiss. She wanted to share some of the CDC's preliminary findings. On the first day of the study back in May, she and the other physicians had sent three residents straight to the emergency room because their blood pressure was so high the doctors feared for their lives. Now, looking at all of the blood-pressure readings together, Kreiss's team could see that high blood pressure was ubiquitous, even among the town's children. The population as a whole had blood-pressure rates 30 percent higher than expected. A full third of the residents had clinically high blood pressure. Her team had also found a dozen people with undiagnosed diabetes, nine with troubling levels of protein in their urine, another nine with signs of serious liver disease, and high cholesterol overall.

But she didn't have their DDT levels yet. They wouldn't be ready for several more months, she said. For the town, the wait was agony. "There is a lot of anxiety and frustration down here," Foster told one of the reporters who came through.

Early in January 1980, Foster got another call from Kreiss. The CDC routinely tested a sample of Americans, about every five years, to get a sense of the population's average bodily levels of pesticides and chemicals, including DDT. The samples her team took from Triana showed that the townspeople's levels were five to fifty times higher than average. Levels were highest, again, in those who ate the most fish. They were also highest—unexpectedly, Kreiss said—in those who were oldest. The long-standing claim that DDT levels reached equilibrium, which Hayes had long testified to in court and other hearings, wasn't true. And people with higher levels of DDT also had higher levels of a liver enzyme, gamma-glutamyl transpeptidase, that helps break down toxic substances in the body. The levels suggested their bodies were working harder to break down their chemical burdens.

Kreiss also shared something else. During lab analyses of the samples, it had become clear that many of the residents' blood contained an entirely different set of chemicals: PCBs. The industrial chemicals had been banned three years earlier, for reasons similar to those behind DDT's ban: PCBs caused cancer in lab animals, persisted in the environment, and magnified up the food chain. In fact, Kreiss told Foster, it was possible that PCBs, not DDT, were behind the town's high blood-pressure and cholesterol levels. She'd need to carry out additional studies to be sure. But right then, it wasn't clear where the PCBs had come from, so the EPA was dispatching scientists to sample the local creek water, sediment, and fish, Kreiss said.

For the town, the news was a painful blow. "We may be the most polluted people in the nation," Foster said.

All the while, reports on the evening news carried footage of President Jimmy Carter addressing the "national disaster" in the Love Canal neighborhood in Niagara Falls, New York. The neighborhood had been built on a former dump site for a company called Hooker Chemical and Plastics Corporation. Now, decades later, old storage drums had broken up through the ground in people's backyards, and chemicals had oozed into their basements. A group of white home owners had demanded that the government evacuate them. A couple of them had even taken two EPA officials hostage. The state had responded by declaring a health emergency, and the Carter administration, in response, pledged federal aid.

Foster, watching the events unfold from Triana, saw his town's anxiety shade into outrage. "We've got all kinds of chemicals in our blood, but nobody is listening to us," he said. "They're too busy listening to the people up at Love Canal."

Foster said as much to every reporter who came through town. He applied for aid from every federal agency he could think of. When a reporter for a civil rights magazine came by, Foster told him something that he just could not get out of his head: that back in 1964, when his town was incorporated and he was named mayor, not just Olin but the US Army and the US Department of the Interior all knew that the creeks and their fish were contaminated with DDT. But no one told them until 1978.

Over the next few months, EPA investigations found that a sewage-treatment lagoon on the arsenal contained a thick layer of PCBs at its bottom. When the lagoon periodically overflowed, the chemicals spilled out and made their way into nearby surface waters. The agency called a town meeting to share the findings; Kreiss was also invited to help explain what it all meant for people's health.

Foster had every intention of remaining civil at the meeting, but he wanted to make sure that the agencies knew how angry everyone in Triana was. So he planned a march. The townspeople made signs, set a time and a gathering place, and called reporters in Huntsville and Birmingham. On the morning of the meeting, they marched through Triana's streets, chanting and clapping. Foster took the lead. Marvelene Freeman, a local grocery store owner and a close family friend, marched with her children to his left. Four hundred people lined up behind them. Foster called it a "frustration march," an ironic twist on the civil rights freedom march.

At the meeting itself, Kreiss attempted to offer reassurance. Although it was true that high blood pressure combined with high cholesterol could increase a person's risk of heart disease, stroke, and kidney problems, there didn't appear to be a connection between people's DDT levels and those conditions. The PCBs in their bodies were an unexpected finding, but the town's levels were not abnormally high and posed no threat to anyone's health, she said. They were surprising only because no one knew, until now, that the fish in Indian Creek and the Tennessee also contained PCBs.

She made a request. The CDC wanted to continue its study, with the town's cooperation, of course, to find out whether—and how quickly—the people were eliminating both chemicals stored in their fat. Such a study could help CDC scientists figure out how best to eliminate heavy chemical burdens in other communities in the future.

The plea went over like a bomb. Foster found himself thinking of what had bothered him the year before: the sense that it wasn't right that so many federal agencies had known that there was DDT in the creek when Triana was incorporated in 1964. Now it seemed that they had been using the town as a group of guinea pigs all along, allowing them to build up chemicals in their body so they could study them later on. It was a blatant example of experimentation on Black people.

The town was not interested in elimination research, Foster told Kreiss. They wanted biopsies, autopsies, and morbidity and mortality studies. They wanted to know, as soon as possible, what deaths and diseases in their community were caused by the chemicals they had been forced to carry within for so long without their knowledge or consent. Kreiss told him that she understood. "You're not going to be guinea pigs," she said.

But the kinds of studies they were asking for would be time-consuming and probably inconclusive. There just weren't enough people in the town to draw firm conclusions about the relationship between their DDT levels and causes of illness or death. An elimination study was both feasible and scientifically valuable. It would also help the townspeople understand what was happening to the chemicals in their bodies over time. She couldn't force them to participate, however; all she could do was ask. "If you don't think our efforts are worth your time," she said, "I want you to let us know." No one objected further.

Chapter 21

ASSESSING RISK

The commercial fishermen sued Olin first, in the summer of 1979, for $500 million. They claimed that the company's pollution had cost them their livelihood because markets had abruptly stopped buying their fish. The state of Alabama sued Olin next, that September, for $3 million. The state's attorney general, Charles Graddick, claimed that the company had violated the Clean Water Act and created a public nuisance. Clyde Foster hadn't intended to sue, but after learning that Triana was contaminated with PCBs in addition to DDT, he started to think differently. He reached out to a small Huntsville law firm, and an attorney there who supported segregationist policies nonetheless agreed to help. But Foster wanted more than he could offer. Foster pressed him to bring in a big, out-of-state firm to work the case. He obliged. Through a law-school connection, he connected Foster with a couple of attorneys at the top-flight firm of Kilpatrick & Cody in Atlanta, which had a history of taking on cases concerning voting rights and the death penalty.

The Kilpatrick & Cody lawyers were interested, but as they looked into the matter, they told Foster that no case alleging DDT's harm to humans had ever been won in a court of law. Foster, undaunted, persisted. Just because no case had proved it harmful, he said, no case had proved it harmless, either. Nor had any case involved people as polluted as those in his town. On top of that, the CDC was looking into their health and had already found problems.

When Kilpatrick & Cody attorney Bob Shields heard that, his ears perked up. When he learned that government scientists with the CDC had *published* a report of health effects in the people of Triana, they perked up even further. That sort of thing never happened. Shields and a fellow partner began to lobby their senior partners to get involved.

In the spring of 1980, Kilpatrick & Cody's partners agreed, on a contingent fee basis, to take Triana's case. The Fosters' family friend, Marvelene Freeman, agreed to serve as lead plaintiff, and in short order Foster convinced five hundred Triana residents to join her. The plaintiffs alleged that Olin had deliberately withheld the information that DDT had entered their food chain, an action that resulted in a long list of health problems for the town: poor calcium formation, reproductive system abnormality, hypertension, elevated liver enzymes, anxiety, and trauma. Many in the town believed they were suffering elevated rates of cancer, too, but the Kilpatrick attorneys held back on that point; they knew it would be nearly impossible to prove.

The plaintiffs sued for $1 million apiece. Kilpatrick & Cody's attorneys decided to ask the judge to set a date for a bellwether trial, which would allow them to present just a handful of cases representative of the plaintiffs, whose numbers now exceeded a thousand. Olin's attorneys, from an Atlanta firm just blocks away, came up with their own strategy, asking for time to conduct discovery proceedings with every plaintiff. Both sides knew such a process could take years.

Later that same year, the US Justice Department also sued Olin, for $3.5 million. The department claimed that the company had violated the Clean Water Act, the Resource and Recovery Act, and the Refuse Act through unlawful discharge into US waters and a federal refuge. Then the EPA sued Olin as well, under the Rivers and Harbors Act of 1899.

Eight days later, on December 11, 1980—forced into action by the events not at Triana but those up at Love Canal—Congress passed the Comprehensive Environmental Response, Compensation and Liability Act, authored by House Democrat Jim Florio of

New Jersey, a state that led the nation in its number of toxic dump sites. The new law directed the EPA to create a ranked list of the most-contaminated sites in the country and oversee their cleanup, paid for by the polluters, with support coming from a $1.6 billion fund generated by general taxes and fees paid by the chemical and petroleum industries. The list was called the National Priorities List. The fund was called Superfund. Love Canal was named the first Superfund site.

That winter, Kilpatrick's attorneys realized that US District Judge Frank McFadden had previously been a partner in the law firm representing Olin. They asked him to step down. McFadden refused. Up in Washington, Alabama congressman Ronnie Flippo, a Democrat from Florence, made sure Representative Florio knew about the small, poor, Black town in his state where the creek was poisoned with DDT. Florio entered Triana's plight into the *Congressional Record* as an example of the type of problem his bill was designed to address.

In the meantime, the Army Corps of Engineers unveiled a proposed cleanup plan for the Tennessee and its tributaries that would involve rerouting several miles of the Huntsville Spring Branch, the Tennessee River tributary closest to the old plant site, and dredging the branch and Indian Creek, at a total cost of $88.9 million.

When reporters asked Foster what he thought of the army's plan, he said it sounded fine. But he didn't want anyone to forget that the problem was bigger than the creek.

"There are two issues—what they will do about the environment and what they will do about our health," he said. "Certainly if we can solve one, we can solve the other."

•

While carrying out the first two studies in Triana, Kreiss had had to learn as much as she could about DDT. She read Hayes's studies and others that had come out of his lab, but she wasn't swayed or impressed by his findings or his insistence that DDT was harmless. While at Harvard she had been part of a team of researchers who had discovered that workers in a Massachusetts foam factory were suffering bladder damage because of a new chemical used in the production process. It left her with the sense that as a scientist, there was always an opportunity to look at something more closely, or differently, and discover something new. In this case, plenty of scientists before her had studied DDT, but most had studied people exposed through work in a factory or on a farm. A "food chain" exposure, she realized, would be much more relevant to anyone living in the US.

With the help of Kimbrough and a few other CDC toxicologists, Kreiss dug back into the literature. She had found a collection of studies examining the health of men who had worked with DDT in factories and in fields. She got a copy of the data from the most recent survey of the US population—the National Health and Nutrition Examination Survey—which had last measured average DDT levels in three thousand randomly selected Americans in 1976. She also tracked down two publications that were considered definitive reports on DDT and its health effects, both published by the World Health Organization over the last few years.

Reading through the WHO reports, Kreiss learned that DDT was widely found in wildlife, stored in the fat tissues of mammals, fish, reptiles, and birds. She learned that it was found throughout the US food supply, although its levels there were dropping. It was also in most agricultural soils, most surface waters that had been tested for it, in drinking water, and in marine animals swimming in oceans far from where it had been used. Fifteen years after DDT was applied, half of it still remained. Another fifteen years later, half of that amount remained. The chemical had been banned from the US and much of Europe for years, but it was making its exit slowly.

Kreiss learned that 99.4 percent of people in the US had measurable amounts of DDT residues in their body. Certain groups had average levels that were higher than other groups. Blacks had higher average levels than whites. Poor people had higher levels than middle-class or wealthy people. People who worked in agriculture had higher levels than people who didn't. In her sample of five hundred people from Triana, most were Black, 83 percent lived within half a mile of a cotton field, and nearly 90 percent had done farmwork. In other words, for all of those reasons they were already likely to have high levels of DDT, Olin's plant notwithstanding. And on top of that, the town's long reliance on the creek for household water—until Foster installed the pumphouse—meant they were also predisposed to higher levels from some of their water use.

Reading on, Kreiss learned that fat tissue took up DDT slowly, which meant that a large, one-time dose didn't always translate into as much DDT storage as small doses over time did. Fat tissue took up DDT fastest at first, then more slowly over time—which might have been why toxicologists like Hayes had long believed it reached a state of equilibrium. When people or animals consumed DDT, their bodies broke some of it down into two metabolites, DDD and DDE. The first was largely excreted in urine. The second was stored in fat. So much of the DDT residue in people's bodies was in the form of DDE, which was more quickly and easily stored in fat than its parent compound. And the compounds' main effect was on the nervous system: doses large enough to kill did so by causing respiratory failure or, in some species, ventricular fibrillation, a deadly acceleration of the heartbeat. DDT's second-most significant effect was on the liver, where it caused lesions and enlargement at first and then, over time, nodules. DDT also seemed to stimulate or inhibit liver enzymes, with its effects differing by enzyme. And one of DDT's isomers, o,p'-DDT, was weakly estrogenic, meaning it acted a little like estrogen in the body. Most experts, though, didn't find this terribly important.

She read more. DDT caused liver tumors in mice, and in some experiments the tumors metastasized. It caused liver tumors in trout, but the published studies didn't compare the trout to adequate numbers of control fish. DDT's metabolites caused lung tumors in mice, but when given to dogs or monkeys or hamsters, neither DDT nor its metabolites caused tumors at all. It was difficult to conclude anything about human cancer risk from such evidence, although a study of terminal cancer patients found that they had higher fat concentrations of DDT than control patients did. Still, it wasn't the type of evidence that confirmed a causal relationship between DDT and cancer.

PCBs were trickier to read up on. No one had published average PCB levels for the American population. But a few unpublished reports did exist, and they suggested that Triana residents' levels were not exceptionally high. As with DDT, their PCB residues seemed to accumulate with age. And as with DDT, they were clearly getting the PCBs from fish. The people with higher PCB levels in Triana were also more likely to have higher blood pressure. The connection seemed important, but without a control group it proved nothing.

Still, her Epi-2 study findings were critical. They deserved a bigger audience than the CDC's weekly report of disease and death patterns, which is where she had published the results of the first twelve blood tests. Relatively few scientists were studying PCBs' effects on health, but her finding that PCBs seemed linked to high blood pressure and blood cholesterol suggested the issue deserved more attention. On top of that, her finding that DDT levels increased with age completely upended decades of scientific understanding about its fate in the body. Plus, despite a long-standing ban, DDT was still building up in some people. She summarized her study methods and results and sent not one but two papers to the venerated *Journal of the American Medical Association*, one on her DDT findings and the other on PCBs. The journal published them both.

As an epidemiologist, she knew that there weren't enough people in Triana to answer big questions, such as whether DDT or PCBs were increasing the risk of cancer or heart disease. On the one hand, cancer was very common; on the other, specific forms could be so rare that it would take a large study population to determine whether a chemical in the environment was causing a real increase in specific forms of the disease—an increase that couldn't be explained by the myriad other carcinogens to which Americans were regularly exposed. The most valuable thing she could do, she reasoned, was follow up on a few things she *could* answer. With better medical equipment, she could clarify the relationship between the chemicals and high blood pressure. With further samples from people, she could learn whether things such as smoking, drinking, breast-feeding, taking certain medications, or eating meat affected the pace at which people's DDT levels fell. The information could potentially help devise ways of helping other people rid their bodies of toxic chemicals—but only, of course, if the town agreed.

•

Mayor Foster's persistence in applying for federal aid was finally paying off. The US Department of Housing and Urban Development granted Triana $250,000 for housing improvements. The Tennessee Valley Authority granted the town $1.5 million to retrain commercial fishers, tutor local high school students, and build a nature trail, a ball field, tennis courts, a playground, and a ten-acre pond to stock with fish. Early in 1981, the work began.

Even better, the lawyers in Atlanta believed that the Triana plaintiffs had a strong case. They were compiling a list of witnesses and a set of arguments, and felt confident about both. Because DDT caused tumors in mice, they planned to argue, it should be regarded as a *potential* carcinogen in people. But because they knew that a cancer link would be near impossible to prove, they planned to argue that DDT posed much broader health risks, demonstrated by the high blood-pressure and liver enzyme findings, which were especially grave for the town's youngest residents, exposed to the chemical during their developmental years.

But there was one thing Foster was not happy about: how things had gone with the CDC. Kreiss was planning her next follow-up study for that May. As she had told them during her visit the previous summer, she planned to take blood-pressure readings again and collect blood and stool samples to see how much DDT—and PCBs—were left in their bodies. The more he thought about it, however, the less inclined he was to cooperate this time around. The people of Triana were desperate to know what the chemicals were doing to their livers, their hearts, and their risk of dying of cancer or some other dread disease. They still wanted autopsies and biopsies. "The only thing we know now," said Foster, "is we have 'x' number of parts that measure in our bodies, period." They wanted, more than anything, to make sense of what those "parts" meant.

The attorneys at Kilpatrick had also gotten Foster thinking. What if Kreiss's follow-up study found less of an effect on their blood pressure, cholesterol, or liver enzymes? What if it showed that their bodies were already rid of the chemicals? What if it turned up something that Olin could use to blame them for their own health problems? The town pulled together a litigation committee to talk through decisions related to the lawsuit. In April 1981, a month before Kreiss was set to come back, the committee decided to call the study off. The study wouldn't answer the questions they wanted answers to, said Foster.

"So hell, we're not interested," he told a reporter.

Kreiss was beyond frustrated. She had told Foster she would do a mortality study when it became feasible, but it just wasn't yet. On top of that, she had been working full-time on the elimination study for months, assembling a team of seventeen people, designing questionnaires, and locating specialized blood-pressure equipment she'd had to order from England to take to Triana that spring. "I can't do everything at once," she said.

She worried that if the study didn't happen soon, it wouldn't happen at all. There was no guarantee that the agency would still be willing to support it six months or a year from then. She tried, once again, to convince the town to participate. "It's not clear," she said, "what resources we will have down the line."

Foster and the litigation committee shared none of Kreiss's concerns. Their looming worry was that the government would spend millions, possibly tens of millions, to clean up the river, leaving them to suffer illnesses and deaths that should have been prevented long before. The chemicals inside them felt like ticking time bombs. No one knew when they would go off or what they would do.

That spring, the Urban Environment Conference invited Foster to Washington for the conference's biannual gala, held to honor people who had done outstanding work on behalf of the urban environment. The conference, founded a decade earlier by Michigan senator Philip Hart, wanted to recognize Foster for his legal battle against Olin. No one seemed to notice, or care, that Triana was not a city. At the gala, Foster moved through the spacious and sleek social hall in his double-breasted navy suit. When the time came to give his acceptance speech, he found himself choked with emotion. "We have the highest levels of DDT and PCB in our blood ever recorded," he told the crowd. "Our kids are not growing to normal height. We have more cancer than normal, rare cancers, like cancer of the liver and spleen." No study had shown as much, but the people of Triana themselves had been keeping a local count. The speech moved the crowd to tears.

But back home in Alabama, Foster's daughter Edith teased him. "What, you're an environmentalist now, too?" she chided.

The label stuck. Foster was invited back to Washington a few months later to testify before the Senate Environment and Public Works Committee. He returned two months after that to testify before a Senate subcommittee. Every time he was asked to speak about his town's environmental contamination, he steered the conversation back toward his people's health. Halfway through 1981, Foster said, Triana had already lost seven people to cancer, another was in the hospital, and two others had been diagnosed with the disease. That was far too many for such a small town. And everyone in town believed it was caused by DDT.

"Having people eat DDT for a generation," he said, "is the same thing as taking a gun to their heads and pulling the trigger."

When he returned from his trips, Edith offered to snip the elastic airline tags off his luggage handle. They looked messy to her as they began to pile up. But he wouldn't let her touch them. Leave them, he told her. They were a sign of his life's work.

•

By this time, five parties had sued Olin over the DDT in the river: the federal government, the state of Alabama, the Whitesburg Commercial Fishermen's Association, the Triana residents, and a separate, smaller group of Triana residents led by James Erskine Parcus, another local grocery owner and the only white person on Triana's city council. In late April 1981, the US District Court of the Northern District of Alabama consolidated the five cases into one and ordered a single discovery file for all of them, to be held in the federal courthouse in Birmingham. Judge McFadden was removed, and Judge Robert B. Propst was installed in his place.

Propst agreed to consider the plaintiffs' request for a bellwether trial, but only once each plaintiff completed an interrogatory from the defendants, a questionnaire used in the discovery process. Shields took the proposal back to the firm. Getting more than a thousand plaintiffs to complete fifty-page questionnaires with legal assistance could take a year or two. But the firm's partners decided to take a gamble. They asked every attorney in the firm to pitch in. With Foster's help, they pulled in a cohort of A&M graduate students to help too. Thirty days later, they triumphantly delivered an eight-foot tall stack of completed questionnaires to Propst. The judge set a trial date for January 1983.

The plaintiffs submitted their amended complaints and requested a mandatory permanent injunction requiring Olin to clean up the polluted waters, restock the refuge with fish and birds, monitor the environment, monitor and study the people's health, and reimburse the US government for all costs associated with time spent investigating, enforcing, and remedying the DDT problem in and near the Redstone Arsenal. It was no small request.

As the parties prepared for trial, Triana grew quiet again. Fewer reporters came calling. Fewer government scientists, engineers, and officials passed through. And the town felt like it was held in a state of suspension. Triana had made it onto the federal Superfund list as part of an eleven-mile contaminated stretch of the Huntsville Spring Branch and Indian Creek tributaries, but cleanup would happen only when the state of Alabama agreed to contribute its share of the costs, and the state was refusing. The TVA-funded lake had been dug, but for the moment it was just a broad, deep pit of red clay. The EPA had confirmed high PCB levels in fish, but the chemicals' origins remained a mystery. And the CDC study was still on hold, the town unhappy with its focus, and Kreiss dispatched to a new assignment in Colorado.

Meanwhile, Foster's rage over the whole situation continued to swell. "We're just like mice in a cage here, but no one is studying us," he told a rare reporter. "You know why? I'll tell you why. Because these are black rats in these cages. . . . These are black rats and that's why no one is interested." Foster's strongest ally was still the TVA, one of whose directors, S. David Freeman, agreed. "They need constant medical attention and they aren't getting it," he said. "In a different city with a different racial mix, there would have been a firestorm of reaction."

As the attorneys for the plaintiffs and defendants prepared for trial, it became clear that Olin planned to argue that the town's health was fine—and that their DDT levels were harmless, especially for their race. They pointed to evidence from the same 1976 survey Kreiss had consulted to argue that because DDT levels in southern Blacks were already 50 percent higher than levels in other Americans, the Triana levels weren't exceptional. Of course the townspeople had more DDT in their bodies "than a Wall Street banker," one Olin attorney said. The company shouldn't shoulder the blame, they held, for something racially ingrained. The argument fit a long-standing pattern of using science to validate racial differences born of oppression.

The company's lawyers also planned to deny DDT's toxicity to people generally—regardless of race. Wayland J. Hayes, now sixty-five and an emeritus professor, was their star witness. In an affidavit he prepared for the company, he wrote that scientific studies published since the EPA's 1972 ban contradicted Ruckelshaus's position on the pesticide. DDT caused tumors, he granted, but only in some rodent species. He dismissed the matter of whether or not they were malignant by arguing that because the tumors were unique to rodents, the whole matter didn't apply to people. Studies of plant workers, he said, found them healthier than other men their age in the general population. "There is no evidence," he affirmed, "that DDT is carcinogenic in man."

Judge Propst ultimately limited both sides to no more than forty-five factual witnesses each and further limited Olin to twenty-two expert witnesses on the plaintiffs' complaints. Olin's attorneys complained that the private plaintiffs' health problems were so broad and ill-defined that there was no way they could assemble an appropriate witness list under such constraints unless they examined the plaintiffs themselves. The firm's attorneys asked the court for permission to take additional blood samples from the plaintiffs. The court granted the request.

The plaintiffs' lawyers, well aware of the controversy over DDT's carcinogenicity, stepped carefully around the matter. They alleged that DDT might increase the likelihood of cancer, other diseases, or diseases yet to be discovered. They sought their own star witness who could testify as much, an Englishman named Ian Nisbet, who was an expert in a relatively new scientific discipline called risk assessment. Nisbet had a science dilettante's résumé: he had studied fluid mechanics and bird flight, had taught at Harvard and the University of Malaya in Kuala Lumpur, and had worked as an engineer and biophysicist. In the late sixties he had become interested in toxicology, and in the early seventies he had become intrigued by the developing field of risk assessment, the holistic consideration of risks that toxic chemicals posed to human health. He had published hundreds of papers and a book, out just then, on how chemical exposures affect human reproduction. He had also studied persistent chemicals' effects on the environment for the EPA and authored a federal report on DDT for the National Institute of Occupational Safety and Health, part of the National Institutes of Health (NIH). Risk assessment was a discipline so new it wasn't even taught anywhere; Nisbet explained it as a complicated application of scientific knowledge from a range of scientific fields—including chemistry, toxicology, pharmacology, epidemiology, and statistics—to determine the full set of hazards that a chemical exposure posed.

DDT was a perfect example of just how complicated risk assessment could be, Nisbet explained. Everyone knew the pesticide was stored in fat. But the form stored in fat was largely DDE, the main result of metabolizing DDT, and DDE didn't just stay in fat indefinitely; it flowed continuously out of the fat, into the bloodstream, and from there to the fat of other organs. On top of that, DDE and DDT's other breakdown product, DDD, each existed in two isomers, p,p' and o,p', or para, para prime and ortho, para prime. And the rate at which DDT's isomers and metabolites moved in and out of fat, and the rate at which the isomers were metabolized, also differed—not just from one isomer to another but from one person to another and one context to another. Diet, exercise, illness, age, and other factors could speed up or slow down metabolism and movement. So in truth there were six chemicals in a "quasi-steady" state acting differently dependent on context. Risk assessment attempted to take all of that, and more, into consideration.

In the case of Triana's DDT exposure, Nisbet considered what type of DDT the people had been exposed to, in what amounts, at what age, and for what duration of time. He then consulted all types of studies, human and animal, whose conditions most closely resembled that of the Triana residents. In his view, there were about five health effects seen in animals that could be expected to appear in the people of Triana because they were so consistently seen in studies or experiments in which the exposure levels were similar to those in Triana: fatty liver, induction of microsomal enzymes in the liver, reproductive harm related to hormonal changes, immune system impairment, and cancer. He pointed to evidence that when DDT and DDD were fed to mice over two generations, the second generation developed more cancer cases than the first generation. Studies like that, he noted, suggested that it was possible that the *timing* of DDT or DDD exposure might affect future cancer risk.

Nisbet would later find himself explaining to Judge Propst that how and when DDT reached an organism and was metabolized by that organism could have unforeseen and potentially dangerous consequences that scientists had yet to bring to light. Judge Propst found Nisbet's testimony so confusing that he repeatedly asked Nisbet to slow down and reexplain things. But three things that Nisbet said were easy to grasp. The first was that DDT might cause cancer; the science wasn't definitive. The second point was that the risk of any unwanted outcome increased with dose, and the people of Triana had extraordinarily high doses. The third was that, as Nisbet had learned from Kreiss's findings, most of the DDT the people of Triana had consumed—two-thirds of it—was DDD.

"Is that good or bad from the standpoint of what effect it has on them?" Propst asked him.

"From the standpoint of the scientists, it is bad," said Nisbet. There were no studies on people exposed to DDD, he said. It was entirely possible that DDD had "some extraordinary toxic effect" and that no one knew it yet.

PART III

Chapter 22

SETTLING

Elizabeth Whelan had a comfortable suburban home in New Jersey, a beach house on the Jersey shore, an apartment in New York City, and an eighteenth-floor office with views of the Hudson River from which she ran her organization, the American Council on Science and Health (ACSH). She had launched ACSH in 1978, when she was thirty-four, because she believed that the country needed a new type of consumer advocacy group, one to remind people that drugs, pesticides, and food additives made for longer, healthier, and more productive lives. She was fiercely devoted to the cause. Her daughter, Christine, always joked that she wasn't an only child because her mom's other child was "Acksha," as they called ACSH.

Whelan—with public health degrees from Yale and Harvard and a bounce of red hair—figured out early on that the best way to get her organization's message across was to become a media star herself. By the early eighties, that's what she was: she had a syndicated radio show, a daily commentary on Ted Turner's new Cable News Network, and two monthly magazine columns. *People* magazine profiled her, shooting pictures of her drinking diet soda and spreading supermarket icing on Christine's birthday cake. Her star rose because she said things that few others with an advanced degree in epidemiology would say to the press: that artificial sweeteners were harmless, that soft drinks belonged in school, that people had a right to eat junk food, and that worries about heart disease, cancer, and "alleged environmental hazards" were all exaggerated.

"Ralph Nader, move over," the *Wall Street Journal* said.

Whelan's star also rose because the nation's biggest corporations needed little convincing to write her checks. By 1981, ACSH was bringing in $750,000 a year from the likes of Stauffer Chemical Company, CIBA-Geigy, Diamond-Shamrock Corporation, Coca-Cola, Frito-Lay, and Hooker Chemical, the company behind Love Canal's toxic waste. From the start, one of her go-to funders was the John M. Olin Foundation, founded by John Merrill Olin back in 1953, before the company merged with Mathieson Chemical Corporation. John's father, Franklin W. Olin, had founded the company that became Olin Chemical Corporation, and John had long served as president and later board chair. For nearly a generation, the John M. Olin Foundation had given money to hospitals, museums, universities, and the like. But in the seventies, frustrated by the nation's liberal turn, Olin had decided to use the foundation to a different end: to fund scholars, thinkers, and opinion makers to support free markets and shift the nation back to the right.

ACSH served the foundation's new aims. It was one of a new breed of think tanks that emerged at the time to sell conservative ideology to politicians and the public. They sowed doubt about settled academic scholarship, cultivated industry-friendly experts, undermined unbiased ones, and served journalists and politicians a steady stream of statistics, talking points, and self-serving facts. It was the same work that Hill & Knowlton and other Manhattan PR firms had pioneered for the chemical industry and the tobacco industries in the fifties and sixties, but the new think tanks did it under the guise of independence. Among them, ACSH had carved out its own niche, focusing uniquely on the science of food, the environment, and health.

By the early eighties, while Olin Chemical was paying its lawyers to defend the company in court, the Olin Foundation was paying Whelan to defend free enterprise in the public sphere. Her organization defended DDT, belittled worries about pesticides in food and in the environment, downplayed the scale of the nation's cancer epidemic, and criticized the science driving chemical regulation. "We don't feel a chemical is guilty until proven innocent," she told reporters repeatedly. And "we don't think a chemical should be banned at the drop of a rat."

•

Kilpatrick & Cody's attorneys spent 1982 preparing eight witnesses: seven adults and a Triana teenager selected by Judge Propst, who would take the stand when the case went to trial early in the new year.

But in the meantime Propst had assigned the case to another federal judge for the northern district of Alabama, James Hancock. Hancock had a reputation for settling complex cases, and late in 1982 he invited the lawyers for the plaintiffs and the defendants to make opening statements in brief. Then he met with them all separately. Everyone knew that a bellwether trial could be disastrous for Olin. If the verdict resulted in large awards for the first handful of plaintiffs, subsequent claims could reach a total cost impossible to meet. John M. Olin himself, long retired, died that fall at his home in East Hampton, New York. The decision on how to protect his father's company would ultimately be left to Olin's board.

In December, Judge Hancock invited the attorneys for both sides to a joint meeting. With everyone assembled, Hancock proposed a settlement under which Olin would pay $24 million to the plaintiffs and another $75 million to $100 million in cleanup fees. "The sum was massive," said Kilpatrick & Cody's Bob Shields. It was also a shock, for settlements that large were rare and Hancock hadn't so much as hinted that he was thinking of such numbers. Hancock turned to each lawyer in the room to ask them whether the terms were suitable. Each one said they'd need to check with their clients. Olin's attorneys said the company would need to put the proposal before the company's board. Shields and the other attorneys at Kilpatrick & Cody held their breath. They placed no bets on what Olin would do. They continued preparing their witnesses in the event that they'd still be going to trial early in the new year.

Then, on Christmas Eve, word came through. Olin was prepared to offer a $24 million settlement to Triana.

Under the terms of the settlement, Olin agreed to put $1 million a year for five years into a health fund for the town and to pay the remaining $19 million to 1,100 Triana residents over the next five years, starting in January 1983. It offered a lump sum of $300,000 to the three dozen commercial fishermen. In exchange for release from US government claims for restitution and related claims made by the state of Alabama, the company agreed to pay the complete cost of an environmental cleanup that would remove all DDT from an eleven-mile stretch of the Tennessee River and its tributaries, and that ensured no fish had more than 5 ppm DDT by the time the cleanup was done. The company would put another $375,000 into a fund to mitigate DDT's effects on other wildlife in the refuge. According to the agreement, called a consent decree, the company would have ten years to complete the cleanup, which would be overseen by the EPA and a six-person review panel.

As word of the settlement spread, reporters returned to Triana. Felix Wynn, still riding his bike through town at the age of eighty-seven, hopped off and did a little jig for one. Not out of joy, but to show the reporter how fit he was. He'd never stopped eating the fish, he said. He didn't mind the posted signs that warned people against fishing in the creek. "I don't feel like it's hurt me," he said.

Clyde Foster reacted differently. If the town's litigation committee agreed to the settlement, after the attorneys took their fees—which they had reduced to $6.8 million from $8.4 million— Triana's adult residents would all receive $9,100 apiece, and the children would receive $10,000 apiece, paid out over the next five years. It was no small sum for a mostly poor community. For a family like Marvelene Freeman's, with five children, that came out to almost $14,000 per year, with the minors' payments put in a trust they could access as adults.

"No monetary compensation can substitute for health," Foster said, but "we feel we have something we can work with."

An Olin spokesperson named Charles Dana said the decision to settle was based on a desire to avoid the costs and "disruption" of years spent on trials and appeals. The company was "convinced" that the DDT posed no health risk to the town but acknowledged that the public saw the situation differently. "Unfortunately," Dana said, "we live in a time when the popular perception regarding a particular chemical too often is inconsistent with the scientific facts." It was precisely the sort of statement Whelan routinely made in the press.

Local papers carried the news of the settlement. So did the *Wall Street Journal* and the *New York Times*, but only because it came the same day as bigger news about the company. Olin, which had sold a major portion of its munitions business the year before, had agreed to buy a majority of the stock of the Philip A. Hunt Chemical Corporation, its first step toward acquiring the manufacturer of chemicals used in semiconductors, circuits, photocopiers, and computer printers. The acquisition, said Olin CEO John M. Henske, was in line with Olin's plans to become a major supplier to the electronics industry. The move would cost Olin $90 million, in a year when the company saw $62.8 million in annual income on $1.823 billion in sales. As it happened, that was a nearly 9 percent drop from its $2 billion in sales for 1981. "The year 1982 was one of the toughest years that Olin has had for a long time," Henske said.

Few, if any, in Triana saw those numbers. At the end of January 1983, the town finally agreed to accept the company's offer.

The first checks, for $2,300 each, came that summer. People cheered as the town's mail carrier, Gregg Barlowe, drew near. Neighbors with checks in hand drove through town honking and yelling. "This is about the happiest day there has ever been in Triana," said a father of ten named Beechel Grays, holding his check in the shade of a hackberry tree. "But I'd give the money back if they could take all the poison out of my body."

Foster was also still worried about his town's health. But now, at least, he had a million dollars in a medical fund and four million more on the way. He imagined the town could use the sum, which seemed so vast at the time, to open a clinic and monitor people's health as they saw fit. The clinic, in particular, was one of his long-held dreams for the town.

•

The EPA held open meetings in Triana over the next few years to share Olin's plans for cleaning up the river. Four hundred towns-people came to the first meeting. Olin presented a plan to dredge the tributaries, which would involve excavating all sediment from the affected waters until no DDT remained. The review panel—composed of four government agency representatives, Foster, and an Olin executive—rejected it. The following year, the panel approved a plan to divert the Huntsville Spring Branch around the affected areas instead. In April 1986, the cleanup work began. Triana held a fish fry to mark the occasion.

Olin's work started with road building; the most heavily contaminated reaches of the tributaries were on such undeveloped land that there was no other way to get equipment and crews to the site. Engineers designed five miles of road. Crews laid them at night, working under floodlights, to avoid disrupting arsenal operations during the day. Once the roads reached the site, the crews cut a 1,600-foot channel into the earth to divert the stream, pulling it away from its historical path and reconnecting it downstream, where the sediment was cleaner. Once the waters were flowing over clean land, the work crews covered the exposed, DDT-laden sediment—nearly 400 tons of it—with heavy-duty fabric, a layer of crushed rock, clean soil, and new plants.

In the meantime, Edith Foster wrote up a staffing plan for the medical clinic. She proposed a full-time family practice physician, a pediatrician, three nurses, a lab tech, and an X-ray tech. Every child would qualify for ten visits in the first two years of life, a full suite of immunizations, and doctor visits and recommended screenings every year. Women would be offered annual mammograms; all adults could be tested for DDT, PCBs, and other chemicals. But what the fund would cover wasn't up to her. Judge Propst held hearings with a selection of experts to settle the question, especially in light of the town's exposures. Kreiss pointed out the need to consider what DDT might do to children who had been exposed in the womb. Hayes insisted there was no reason to expect any injury or harm as a result of the DDT exposure. A court-appointed guardian for the minor plaintiffs, Florence attorney Robert Potts, agreed.

Potts, who followed and reviewed the case closely, found it "unbelievable" that Olin had agreed to pay the town such an enormous sum given that no "clinically discernable damage" had occurred or seemed likely to. The best that could be said, Potts noted, was that DDT contamination had taken a terrible economic toll on the already poor town and that it had caused its residents limitless fear, anxiety, and mental anguish. But his task wasn't to weigh in on the question. It was to make sure that the minor plaintiffs understood the settlement and what it meant for them, from their potential for future health problems to their access, at age nineteen, to funds disbursed to them over the subsequent five years.

Potts called a young man named Maurice, a high school student who liked to play basketball. He took down the boy's medical history: tooth decay, the flu, and chest pain sometimes when he ran. "Once you accept this $10,000, you couldn't sue again," Potts explained. "Do you feel it's fair and reasonable for you, Maurice?"

"Yes," Maurice said.

Potts hung up and called the next minor, and then the next. He kept calling until he had spoken to all 430 of them. He later remarked that he found them all "quite likeable."

•

Judge Propst ultimately ordered that the $5 million health fund cover health monitoring and health care costs for 1,400 total plaintiffs. He created a ten-member board to set the terms and manage the fund. Four board members came from federal agencies: the EPA, the TVA, FEMA, and the CDC's parent agency, the Department of Health and Human Services. Kreiss, who had left the CDC, recommended an environmental epidemiologist she knew at the National Cancer Institute named Tom Mason, whom she thought of as "a former hippie" like her. The other five board members were local citizens: Foster, former police chief Joe Fletcher, grocery owner James Parcus, a commercial fisherman named Max Turner, and Marvelene Freeman, who was named board claims clerk.

The board's first task was to elect a president. Mason flew down from Washington, landed in Huntsville, and drove past the farms and fields that lined the roads to Triana. It was August and "hot as all hell," he said. The board members convened in Triana's town hall to hold a vote and begin making their plans. Hundreds of people gathered in the heat outside, waiting. Some were curious and eager to learn what the board would decide to do. But some were belligerent and angry. Some were drunk angry. Some forced their way in and had to be physically removed from city hall while the board met. They didn't want a health fund. They wanted the money. They believed it was their due, a way to recoup the legal fees that had gone to the attorneys.

While people protested outside, inside city hall the board elected Foster as its president. From that day forward, he set a tone for the board's meetings, which took place once a month for the next two years. Before each meeting, Foster had the fire trucks moved out of the fire station to make room for the public. The board met in private and then with all of the plaintiffs, to listen to their concerns. Some worried the fund would focus on academic studies over health care for the people. Some opposed Foster's idea of a clinic. They wanted to be able to use their own doctors and not be forced to use a town doctor. Gradually, Foster himself relinquished the idea. By Propst's order, the fund covered plaintiffs far beyond the Triana police district, some of them across the river in Morgan County, some of them a dozen miles from town. With a coverage area that big, he knew there was only so much a clinic would be able to do for his town.

Foster eventually floated another idea, one he borrowed from the Osage Nation, whose members all had a "headright"—a right to a portion of funds from the sale of oil, gas, and minerals taken from the million-plus acres of land that made up the Osage Reservation in present-day Oklahoma. Foster noted that every plaintiff covered by the fund had a right to some portion of the fund. He proposed a system in which each plaintiff could use their portion for court-approved health care expenses, which included anything from new glasses to an insurance premium. Whatever an individual plaintiff didn't use in a given year could be held in their account. When they died, any funds remaining in their account could pass to their heirs. It would be a headright to health care.

In 1985, after nearly two years of meetings, the board approved the plan. Foster hosted a fish fry to celebrate. "Any of this fish come from that river?" Mason asked him. Foster laughed and said he couldn't answer.

The angry plaintiffs, meanwhile, stayed angry. They complained that the board had wasted too much of their money already on paperwork and meetings. In a letter to Judge Propst, a group of them called the board's headright plan a "rip off." The attorneys were all paid up front, they complained, while they were now forced to wait for installments and reimbursements.

But to outsiders, some money was better than no money. And as news of the settlement payments and the Triana Area Medical Fund, as it was called, continued to appear in local papers, residents in nearby areas thought it only fair that the compensation reach them, too. So they filed suit. More than a thousand residents of the downstream town of Decatur sued Olin for $2 million. Then five thousand Huntsville residents sued Olin for more than $1 billion. Before long, a total of thirteen thousand people living in the areas around Triana had filed or joined suits against the company.

This time, a US district court judge appointed a special master, University of Alabama law professor Francis E. McGovern, to devise a plan to manage the cases. McGovern consolidated seven suits into a single case. In May 1986, Olin settled them all for $15 million.

McGovern's next task was to devise a fair way to disburse the money. Using an updated version of the national survey Kreiss had used, he decided to award each qualifying plaintiff $500 for each multiple that their DDT level exceeded the DDT level expected for their age, sex, and race. People who could provide a doctor's sworn affidavit that they had suffered from cancer, reproductive problems, central nervous system problems, or hypertension were awarded additional money. The evidence connecting DDT to such health problems was far from settled, but McGovern selected the conditions based on input from medical experts. Plaintiffs with cancer received the highest additional payments, up to $60,000, and those with hypertension received the lowest, at $7,500.

To claim funds, present or past residents of six northern Alabama counties surrounding Triana had to opt in, disclose their illnesses and injuries, and submit to a DDT blood test. More than 9,400 people qualified for compensation. Two of them had DDT levels 40 times above normal. But some people with low readings were so upset that McGovern believed that he had to remind everyone that receiving no compensation was "good news," he said, because it meant a person had "a normal level of DDT."

Years later, McGovern acknowledged a fundamental problem with his distribution plan. The national health survey data he used reflected "the reality of life in the South," he said, where certain ages, sexes, and races had been disproportionately exposed to DDT. In effect, a Black plaintiff whose elevated DDT levels were a result of Olin's dumping might not be awarded as much as a white plaintiff because average DDT levels were higher for Blacks than for whites. McGovern only implied this; he never explicitly stated the plan's effects on any specific race. However, he did say that the aim of his compensation plan "was not to rectify all previous wrongs, but to take the world as it was found at the time of the lawsuit and allocate funds accordingly." The plan thus became one more nail holding racial disparities firmly in place.

•

The original settlement money went fast. Edith Foster was completing her medical internship in pediatrics at a hospital in Florida, and she spent it all on housing and living expenses. Back home, many people bought new cars. Others spent the money on house payments, overdue bills, and groceries. The last payments, the smallest ones, came in 1987. Before long, they were spent, too.

The DDT, noted Marvelene Freeman, was still in the creek. "Nothing has changed here in Triana," she told a reporter for *ABC News* who came down with a cameraperson. Personally, she said, she believed that the town had been "cheated."

•

Up on the arsenal, the cleanup crews continued their work, diverting, burying, and monitoring. By late 1987, Olin's operations had removed 379 tons of DDT from the river's tributaries. The following year, DDT levels were down to 89 ppm in Indian Creek smallmouth buffalo, 36 ppm in catfish, 20 ppm in bigmouth buffalo, and 13 ppm in white bass. Levels in brown bullhead had mysteriously doubled, to 120 ppm. The fish still had a long way to go.

Nonetheless, the National Wildlife Federation, party to the suit that the EDF had brought against Olin in 1970, twice nominated the company for its annual "corporate conservation" award. Olin didn't get the award, but the EPA commended the company for meeting every deadline in the consent decree and for largely cleaning up in six years what it had poisoned over twenty-three.

The DDT was still in people's bodies, of course. And the Medical Fund board was still meeting once a month, although now it convened in a hotel in Huntsville. The board was reimbursing people bit by bit, but it had also invested some of the first million, and some of the second million, and so on. The money was growing.

So was suspicion. In some quarters, far from Triana, new evidence was pointing more directly than ever at a relationship between a specific type of cancer and exposure to DDT. And Elizabeth Whelan stood ready to dismiss it.

Chapter 23

HAND-ME-DOWN POISONS

The Wingspread Conference Center sat on thirty-six acres of prairie and woods on the shore of Lake Michigan in Wisconsin, a graceful and sweeping Frank Lloyd Wright building at its heart. It was July 1991, and the surrounding prairie was lush, the nearby woods thrumming with life. When Theo Colborn got there, the weather was cooler than usual, topping out at 71 degrees. She greeted the scientists she had invited as they arrived. Frederick vom Saal flew in from Columbia, Missouri. Ana Soto flew in from Boston, Melissa Hines from Los Angeles, Howard Bern from Berkeley, John McLachlan from North Carolina's Research Triangle Park, and more than a dozen others from all over the US and parts of Canada.

The scientists hadn't ever worked together, for the most part, and most of them didn't even know of the others. But as Colborn had discovered, they all had something critical in common. They were all experts on the endocrine system, in people or in wildlife. And they had all found, in their own fieldwork or lab research, that the endocrine system, a collection of organs and the hormones they secrete, could be knocked out of balance by even small amounts of certain chemicals now found widely in the environment. That was why Colborn, a genial and energetic sixty-four-year-old with oversized glasses and a pixie cut, had invited them all to Wingspread to meet.

Colborn had found the scientists while looking for a pattern, but the pattern in their research was not what she'd been expecting to find. After working as a pharmacist in New Jersey for decades and then running a Colorado farm with her husband and four children, she had gone back to school for a PhD in zoology at the University of Wisconsin. She finished the degree at age fifty-eight. Her children were grown. She accepted a position as a Fellow at the US Office of Technology Assessment and moved to Washington, DC, to write reports on water and air pollution for Congress. As her two-year fellowship was ending, a man named Rich Liroff called her from the merged organization World Wildlife Fund and Conservation Foundation. The foundation needed someone to write a detailed report on the status of pollution in the Great Lakes. She said yes. While doing her PhD, she had taught classes on limnology, the study of lakes and other bodies of fresh water, and she had always been curious about why Great Lakes fish populations had so dramatically declined. Now she could figure it out.

Colborn had begun the work by collecting all of the reports and scientific papers she could find on the lakes and their wildlife, and then sorting the thousands of documents she found into categories, by chemical or species. When she started, she had been sure that she'd find a pattern of elevated cancer rates among wildlife. But the pattern just wasn't there. She decided to look instead at just the reports on wildlife species that had suffered the most problems—cancer in some cases, but also wasting, thyroid problems, unusual behaviors, and abnormalities in their reproductive organs.

That was when she noticed something.

The affected wildlife were all predator species: bald eagle, herring gull, mink, otter, others. Then she noticed something else: it wasn't the adults but their offspring that suffered the problems. And the same chemicals kept popping up across the papers: DDT, dieldrin, chlordane, lindane, PCBs, and others. They were all chemicals known for their environmental persistence.

Tentatively at first, Colborn had developed a hypothesis. The lakes seemed clean, but low levels of persistent chemicals were still biomagnifying up the food chain. Thanks to bans and new EPA regulations, the levels of those chemicals were no longer high enough to cause the dramatic reproductive failures seen in earlier decades, but

Colborn wondered whether the now-low levels were behind the host of problems so many scientists were documenting: male birds with ovarian tissue, female birds with excess reproductive tissue, and fish with male and female reproductive organs, among countless other effects. Because the effects were mostly manifested in offspring, she thought of the chemicals as hand-me-down poisons.

When she had handed her report to Liroff, she felt like she had just scratched the surface of something enormous. She applied for funding to keep the research going, and two Canadian agencies gave her grants. She used them to pay her salary from the Conservation Foundation, which was then absorbed into the World Wildlife Fund. Then one day, she went to hear a talk by Audubon Society scientist John Peterson Myers about his research on sanderlings in coastal Peru. The talk set her on course for the next ten years.

The birds Myers studied, which migrated from South America to the northern US coasts and Canada in a five-day nonstop flight, were in dramatic decline. The birds fed voraciously before their migration, doubling their weight over the few weeks before they took off. Driving up and down the coast of Peru, Myers saw that the birds often fed and bathed at the mouths of rivers that carried agricultural runoff into the sea. Banding and tracking the birds with colleagues, he found that at the other end of their journey, the birds had high levels of pesticides in their brains. He hypothesized that in flight, while burning through their stores of fat, the birds metabolized the pesticides, which entered their bloodstreams, reached their brains, and disrupted their sense of orientation, sending them out to sea to die.

Colborn, who had never been shy, went straight up to Myers after his talk and took him by the shoulders. We have to work together, she said. She told him about the hypothesis that had emerged from her Great Lakes research, and he listened, fascinated. Myers was a field ecologist by training, but he was also the sort of person who captivated audiences and set things in motion. As it happened, he had just accepted a position as director of the W. Alton Jones Foundation, a Virginia-based philanthropy founded by an oil mogul and now devoted to protecting the environment. In his new post, Myers had two senior fellow positions to fill and forty million dollars to fund environmental causes. He offered Colborn one of the positions and said he would fund her to turn her findings into a book.

Colborn had jumped at this chance, too, but she knew that first she wanted to assemble the scientists whose papers had helped her piece together this new understanding of the risks that chemicals posed to the health of wildlife—and potentially people. Myers agreed. Another foundation also put in funding. She titled the invitation-only conference "Chemically-Induced Alterations in Sexual Development: The Wildlife/Human Connection."

And now it was the first morning of the conference's meeting at Wingspread. Colborn waited in the lobby, excited and at the same time nervous that nothing would come of the whole thing. Then Howard A. Bern arrived. The Berkeley scientist, bald with dark arched eyebrows that gave him a permanent look of surprise, was one of the world's most revered endocrinologists. He had coauthored the top textbook in his field. When he got to Wingspread, he didn't walk into the lobby; he rushed. "Where's Theo Colborn?" he said to no one and everyone. He had read a recent paper of hers that argued for a new way of understanding the risks posed by chemicals in the environment. "I can't believe it," he said. "This paper lays it all out and it's all there." He reached out to shake her hand and then pulled her in and hugged her instead. Colborn's optimism swelled.

Over the next few days, the scientists took turns explaining their work and their findings to the group. The wildlife scientists went first, sharing slides of bird and fish reproductive organs harmed by chemical exposures. They told of Great Lakes salmon that failed to develop secondary sexual characteristics, female snails on Long Island with penises and penile ducts, and male beluga whales with female sex organs and deficient immune systems. The researchers who worked on humans exclaimed aloud as they noticed parallels to their own work. To Colborn, it felt like the room itself was coming to a boil.

The human health researchers presented next. Bern shared the story of the drug diethylstilbestrol (DES), which was prescribed to pregnant women for twenty-five years before researchers like him found that it was causing a rare form of vaginal cancer in their daughters, exposed in the womb. McLachlan, a government toxicologist, described how he had induced the same cancer in the lab by giving DES to pregnant mice; in the mice's male offspring, the chemical caused lower sperm counts and undescended testes.

Soto, a biologist at Tufts University, told the chilling story of a lab experiment gone wrong. She and a colleague were studying how estrogen causes breast cancer cells to multiply. Then one day all of their cell samples began proliferating, even those they had taken great pains to keep estrogen-free. Something was contaminating the samples. It took them four months to figure out that the contaminant was coming from plastic tubes they used in the lab. Another twelve months later, with no help from the tubes' manufacturer, they figured out that the culprit was a chemical called nonylphenol that was used in the tubes' production. Nonylphenol, also used in detergents and present in hair and skin products, had leached out of the plastic and exerted estrogen-like activity on the cancer cells.

At the end of the first day of the meeting, a group of the scientists stayed up late talking. They talked and drank until they emptied the bar, struck by what they could not ignore. By the end of the third day of the conference, they agreed to work together to draft a statement on the unrecognized hazards of hormone-mimicking synthetic chemicals found throughout the environment. Myers made a suggestion. He had watched how scientists had come together just a year before to issue the world's first consensus statement on global climate change, which of course had tremendous implications for the world's birds. At the Audubon Society, he had written a report in which he used the phrase *climate disruption* to capture the havoc wrought by the warming of the Earth. He thought the term was just as fitting here. He suggested they call the hormone mimics they had found endocrine "disruptors." The scientists agreed. It was a term that would persist, like some of the chemicals themselves.

As the group crafted their statement, they borrowed a framework used by the scientists on the Intergovernmental Panel on Climate Change. "We are certain," they began, with words that echoed the climate-change group's statement, that "a large number of man-made chemicals that have been released into the environment, as well as a few natural ones, have the potential to disrupt the endocrine system of animals, including humans."

"We estimate with confidence," they went on, that developmental problems seen in people today resulted from their parents' exposure to synthetic hormone disruptors in the environment, and that more research focused on the matter would reveal more parallels between impairments in wildlife and humans.

"Current models predict," they continued, that "any perturbation of the endocrine system of a developing organism may alter the development of that organism." Even small changes in hormone levels during particular windows of time in development could change "sex-related characteristics" in irreversible ways.

Science is fraught with uncertainties, and findings must be replicated over and over again before they're accepted as fact. But these conclusions were things they all agreed on. A footnote on the first page of the statement named the synthetic chemicals of concern. DDT topped the list. And with that, the embattled pesticide joined a brand-new class of chemicals.

Princeton's scientific press agreed to publish their statement, collected in a 430-page volume with all of their presented papers. The editor, convinced of the dangers of chlorinated compounds, asked Colborn to submit the text on paper that was chlorine-free. Dozens of scientists reviewed it for them, including Ian Nisbet—the risk-assessment expert who testified on behalf of Triana—now running his own scientific consulting firm from Cape Cod. A scientific journal published a much more concise version, written by Colborn, Soto, and vom Saal. And then someone from the Pew Charitable Trust reached out to Colborn, asking if she would be willing to summarize their findings in a book for the public.

Colborn agreed once again, but as she tried to turn dense laboratory science into engaging and accessible prose, she struggled. As she shopped her chapters around, agent after agent turned her down. Her writing was too dry and technical. She turned to Myers for help. He was happy to try to liven up the text, but the two of them still believed that they needed someone who was a better writer than either of them. Myers suggested a science writer he knew, Dianne Dumanoski, who covered the environment for the *Boston Globe*.

•

When Dianne Dumanoski had first started working for the *Globe* in the seventies, the paper, family run for a hundred years, was flush with money. The *Globe*'s owners, the Taylors ("old-school Yankees" by reputation), cared about the environment, and as a result, Dumanoski's editors spared no expense when assigning her stories. They sent her to scientific meetings around the world. They let her spend a whole week learning about the ozone layer before she had to write a single word. By the early nineties, she had covered deforestation, the spread of acid rain, and the growth of a hole in the ozone layer. She had written about whale deaths and rebounding bald eagle populations. She had followed the emerging story on global warming. The paper was on a crusade, and she was part of it.

But writing about endocrine disruption was another matter. She learned about the theory and its architect, Colborn, from Pete Myers. She didn't know him well, but she had been running into him at environmental conferences for years. Each time they saw each other, dashing from one conference session to another, they had intense conversations that ended too soon, with both of them hurrying off in different directions. On a tip from Myers, Dumanoski had attended a talk that Colborn gave at Tufts University. She listened, rapt, and then tried to write an article—but the paper didn't go for it. At the time, newsworthy stories about pesticides all focused on cancer, and this wasn't a cancer story. On top of that, the whole notion of endocrine disruption was complex and counterintuitive. "I couldn't convince an editor that it was worth talking about," Dumanoski said.

Things were also shifting at the paper. The owners were preparing to sell it. Budgets seemed to be shrinking. Her editors started nudging her away from big, global environmental stories, such as the warming of the Earth's atmosphere; now they wanted local and regional stories, like stories about puffins on the coast of Maine. She decided to apply for a fellowship at the University of Colorado,

Boulder, and when she got it, she took a leave from the *Globe* and went out west for a semester.

She was at a conference of environmental writers at Duke University, held during her leave, when she caught sight of Colborn walking toward her. Dumanoski worried; she was certain that Colborn was going to press her to write about endocrine disruption for the *Globe* and that she'd have to explain that her editors just weren't interested. But Colborn asked her something else entirely, without preamble: she asked Dumanoski to write a book with her and Myers. It was completely out of the blue. Dumanoski asked if they could take some time to talk it over. They found an elegant sitting room, settled into a couch, and Dumanoski listened to Colborn again lay out the science behind the theory and what it all meant.

But writing a book with two people she didn't know was not a decision she was going to make on the spot. It also didn't strike her as a particularly wise decision. She hardly knew Myers and Colborn, and she had never written about endocrine disruption. She needed time to think it all through. Back in Boulder, she talked it over with her husband and friends. Her fellowship would be up at the end of the fall semester, and she wasn't terribly keen on returning to the *Globe* and facing the changes at the paper. Plus, she had a feeling about Colborn. "I think she's really onto something," Dumanoski thought. "She could be the next Rachel Carson. Except she can't write."

At the end of the semester, Dumanoski packed up her things to head home to Massachusetts. A storm chased her as she drove east. She stopped at the Cornhusker Hotel in Lincoln, Nebraska, en route, went to the bar, and ordered a glass of chardonnay. It was her birthday. She had already made up her mind. She was going to help.

Chapter 24

NESTED STUDY

Mary Wolff first came across what was now called the Wingspread statement in the scientific journal *Environmental Health Perspectives*, routine reading for people in her field. She wasn't a wildlife scientist by any stretch. She was a chemist in the Division of Environmental and Occupational Medicine at Mount Sinai School of Medicine, where she had been for nearly two decades by then, studying how industrial chemicals affected people's health.

She was at Mount Sinai thanks to a man named Irving Selikoff, who had hired her back in 1974, a few years after she had finished her PhD in chemistry at Yale. At the time, her classmates were taking jobs at universities or in industry, and she hadn't wanted to do either. She took a job as a chemist for the New York Transit Authority instead, to help the agency reduce emissions from its buses. Then she read an article in the *New Yorker* that mentioned Selikoff and labeled him a "pioneer" for uncovering the link between asbestos and cancer. He was as much of a celebrity as you could find in the field of occupational health. Within a year, she was working for him. The asbestos discovery had left him beloved among labor unions, which helped fund him to study a long list of workplace chemical exposures. Working for him, Wolff never ran out of fascinating and important questions to test in the lab.

Now, reading the Wingspread statement, Wolff's eyes fell on a table on the second page. It stuck with her. It contained a list of endocrine-disrupting chemicals and their effects on wildlife: thyroid dysfunction, decreased fertility, birth defects, abnormal development of the reproductive organs, disruption of reproductive hormones, and more. According to the paper, many of the chemicals had similar effects in animal species and humans. Several of the chemicals were persistent compounds that she had studied on her own and with Selikoff, including PCBs and polybrominated biphenyls (PBBs), widely used flame retardants that they found concentrated in people who lived in areas where the meat and dairy supplies were contaminated. In some of their studies they had also measured DDT, which was listed in the Wingspread table, too.

The table also stuck with her for another reason. When she saw it, she recalled a study she had seen a few years before by Walter Rogan, an epidemiologist at the National Institute of Environmental Health Services in Research Triangle Park. Rogan and his colleagues had collected samples of breast milk from women across North Carolina, approaching them on hospital tours, in Lamaze classes, and at prenatal clinics. Rogan knew that PCBs and DDT were stored in human fat and in nearly every woman's breast milk; a national survey published a few years earlier had shown as much. But no one knew what those levels looked like over the course of lactation.

Rogan's team showed that the chemical levels were highest just after birth and fell substantially as the women breast-fed. New mothers, that is, offloaded the chemicals into their newborn infants, and no one knew what that meant for the babies' future health. But Wolff took something else away from Rogan's study. Chemical levels were lower in women who had already breast-fed a previous child and those who had spent more time breast-feeding. Studies of workers' DDT exposure had often focused on their liver function or neurological damage, but lactation was controlled by hormones. Could the chemicals, acting like hormones, be affecting lactation in some way?

Wolff thought about this in light of something she had worked on herself just recently. As a lab chemist, she was often asked to analyze samples for other scientists' epidemiological studies, and two years earlier her lab had analyzed breast-tissue samples for a team of scientists comparing levels of DDT, PCBs, and other chemicals in a small group of women, some with cancer and some with benign lumps. The levels, they found, were up to 60 percent higher in the women with disease. The study findings suggested the chemicals might play a role in the development of breast cancer. And breast cancer was known to be linked to hormones, too. But if there was a connection there, someone would need to conduct a much bigger study to prove it.

Wolff was talking about just that with some of the other scientists and physicians in her department when one of them told her about a research team downtown, at New York University, that had launched a large cohort study focused on breast cancer. It was what was called a prospective study: it recruited people, in this case all women, and then followed them over time to see what, if any, health problems they developed; it then looked back in time at their exposures to attempt to determine why. The study's lead epidemiologist, an Italian physician-scientist named Paolo Toniolo, had collected multiple blood samples from more than fourteen thousand New York women over the previous several years. So far, seventy-seven of them had developed cancer. Wolff went downtown to meet with Toniolo and talk through her idea: what if they designed a much larger, and therefore statistically stronger, version of the study that her team had just worked on? If he supplied the samples, her lab could analyze them for DDT and PCBs. Toniolo agreed.

With the help of a few other colleagues, Wolff and Toniolo designed something called a "nested case-control" study. They identified fifty-eight women who had developed breast cancer within six months of entering Toniolo's study. For each woman with breast cancer, or "case," they randomly selected "matched controls" from the cohort; the controls were women without cancer who were roughly the same age and had the same number of blood draws and menstrual history as the cases. Then Wolff's lab analyzed the samples, which were coded so that her team had no idea which ones had come from patients with cancer and which ones hadn't. When they were done, they "unblinded" the results. The cases had

PCB levels 15 percent higher and DDE levels 35 percent higher than the controls. The women with the highest levels of DDE— they measured the metabolite because Rogan had and because it constituted the majority of the DDT residues in people—had four times the breast cancer risk of those with the lowest levels. And when they took women's lactation history into account, they found that breast cancer risk fell with each additional month a woman spent breast-feeding.

"These findings suggest that environmental chemical contamination with organochlorine residues may be an important etiologic factor in breast cancer," they wrote in an article published in the *Journal of the National Cancer Institute*. "Given the widespread dissemination of organochlorine insecticides in the environment and the food chain, the implications are far-reaching for public health. . . ."

Wolff and Toniolo knew that the findings were going to cause a stir, but Wolff wasn't quite prepared for just how big the news was. The study was published in 1993, in the midst of a tidal wave of breast cancer activism across the United States. In cities from Boston to San Francisco, women with breast cancer were forming advocacy groups that were demanding a shift in thinking about the disease. Breast cancer had long been a taboo topic, a disease ascribed to women's diet, mental state, or lifestyle choices, one its patients were forced to endure in silent isolation. Now breast cancer activists were insisting that the disease was not their fault. And many of them suspected that the real cause was chemicals in the environment.

Breast cancer activists across the country were demanding that government and academic scientists look more closely at the disease's relationship to chemicals in their food, neighborhoods, workplaces, and bodies. They sent hundreds of thousands of letters to Congress and the White House. They met with the head of the National Cancer Institute. They convinced Congress to double its spending on research. They won seats on federal grant-review panels. One group convinced the CDC to create a panel to investigate the link between breast cancer and environmental chemicals, and then pressed Congress to fund research into the disease's causes on Long Island, where rates were exceptionally high and where many women believed that chemical exposures might explain why. Congress conceded and funded the study at a cost of $32 million.

Some people said the women borrowed all their tactics from the highly visible AIDS activists of the time. Breast cancer activists themselves offered a different explanation for their successes. "We're the activists from the sixties and seventies," said the president of the National Breast Cancer Coalition. "We're used to seeking out and demanding change."

In the midst of this national fervor over breast cancer, Wolff and Toniolo held a press conference on their study findings. "It was huge," Wolff said. They were mobbed by reporters. Wolff talked to journalists from the Associated Press, the United Press, the *Wall Street Journal*, and others, and went on network news and CNN. Headlines the next day announced that breast cancer was now linked to DDT. In the weeks and months that followed, magazines devoted whole issues and television talk shows devoted whole episodes to the relationship between breast cancer and chemicals in the environment.

All across the country, women involved in the breast cancer movement absorbed the news of Wolff and Toniolo's study. Some recalled dousing cows with DDT when they were young. Others recalled running in its mist. Some worried over what they were passing on to their daughters through the chemicals they carried in their bodies. "Those of us who are now in our 40s and 50s," said Ellen Silbergeld, a scientist with the EDF, "are carrying the legacy" of DDT's reckless use decades before.

In Washington the National Cancer Institute was getting relentless pressure from activists. On the news of Wolff's findings, one of its scientists, Susan Sieber, spoke to the press. "We're behind in our understanding of environmental causes of breast cancer," she said. She told reporters that the institute had already proposed a series of studies to investigate the problem. With her own colleagues, she took stock of what they could do quickly to jump-start such research. Her supervisor, Tom Mason, had recently taken another epidemiologist, Bob Hoover, down to Alabama with him. Together, Sieber and Hoover came up with an idea: a study of the women in the Triana Medical Fund.

Such a study might also do double duty. Eighty percent of the women in the NYU cohort were white. That wasn't unusual; largely white cohorts were common among the few studies that had looked at the link so far. But it limited scientists' ability to draw conclusions that would apply to everyone who had been exposed.

•

In some ways, Wolff and Toniolo's study wasn't an anomaly for its time. In the early nineties, it was part of a growing body of research drawing conclusions about the link between DDT and various forms of cancer. Epidemiological studies were linking DDT exposure to soft-tissue sarcomas, non-Hodgkin's lymphoma, leukemia, and other diseases in farmers and factory workers. Not long before Wolff and Toniolo's study was published, in fact, researchers at the University of Michigan and the University of Southern California had published results of a nested case-control study of pancreatic cancer among chemical manufacturing workers, all of them men. The study found a strong relationship between exposure to DDT and cancer; it also found that the longer a man had been exposed to DDT, the higher his cancer risk.

That study prompted headlines too—and a press conference organized in part by Elizabeth Whelan's ACSH, which invited Thomas Jukes and his friend J. Gordon Edwards to Washington, DC, to talk to reporters about DDT's benefits. DDT's ban, said Edwards, repeating one of Jukes's favored phrases from the seventies, was behind an "unholy genocide in underdeveloped countries throughout the tropics." It had been twenty years since DDT's ban, and he and Jukes were still pressing for reversal.

On the whole, ACSH and its members weren't terribly concerned with DDT per se, but they did have a few things to say about the latest research on it. After his study with Wolff came out, Toniolo traveled to France for a sabbatical at IARC, where he found himself face-to-face with ACSH scientific advisory board member Bruce Ames. The Berkeley professor, a decorated biochemist who had previously been on the board of directors of the National Cancer Institute, was renowned for having developed an easy and low-cost lab test for identifying chemical carcinogens. He was also highly skeptical of much of the research linking cancer to synthetic chemicals. Toniolo spoke with him. Ames knew about his well-publicized study with Wolff. Ames "wasn't impressed a bit," Toniolo said. "Quite the contrary, he expressed the frank opinion that we were completely wrong."

•

Out in Southern California, a grueling lawsuit over DDT's environmental wreckage was just moving through the courts. About a decade before Wolff and Toniolo had published their results, wildlife biologists had begun trying to bring bald eagles back to the Channel Islands, an undeveloped archipelago off the Southern California coast. Their efforts failed—and were still failing. The birds they introduced laid eggs, but their shells were always too thin; they broke well before they could hatch. DDT, they eventually learned, was still making its way into the birds' diet, long after the ban. And that meant fish and other marine species were contaminated, too.

The news had given Allan Chartrand, a scientist newly hired by the Los Angeles Regional Water Quality Control Board, an idea. With the world's biggest DDT producer located in Los Angeles County, it was no surprise that wildlife in the Southern California bight were contaminated. The Montrose DDT plant in Torrance had ceased production in the early 1980s, but Chartrand had heard rumors that the company also used to dump its waste right in the Pacific. He asked his supervisor for permission to go down to the plant and ask the Montrose management team if the story was true. His supervisor said yes and dispatched Chartrand and a couple of colleagues.

To Chartrand's "eternal surprise," he later said, company managers were more than willing to answer. They told him and his colleagues that they had sent barrels of acid sludge waste out to sea for decades. When the county asked them to obtain a permit to keep dumping, they did. Their disposal firm, California Salvage, had dumped much of their waste at two authorized ocean dump sites far offshore, north and northeast of Santa Catalina Island, one of the Channel Islands. Prohibited from leaving waste afloat lest it obstruct ships, the disposal workers had punctured the barrels to ensure that they sank. The permit had also required Montrose to log the type and amount of waste it sent off. In a conference room at the Torrance plant, Chartrand said he and his colleagues pored through the logs and saw photos of men onboard barges, taking barrels to sea.

Based on the company's records—and the fact that the acid sludge produced in DDT manufacture was between 0.5 and 1 percent DDT—the water board scientists calculated that there was likely as much as 700 metric tons of DDT in the ocean by 1961. Not all of it was at the authorized dump sites; on at least a few occasions, Cal Salvage reportedly dropped its hauls short of the sites. Chartrand and his colleagues also estimated that another 1,800 tons of DDT went into the ocean closer to shore, discharged by the sewage plant. They borrowed a research vessel owned by the county and took it out to the ocean dump sites to trawl the ocean floor. They pulled up sediment, debris, and "tar cakes," by-products of organic material that form during petroleum extraction. They took them to an institute at UCLA for analysis.

UCLA geochemist M. Indira Venkatesan was a natural collaborator for the water board because she had been measuring DDT and PCBs in the bight's water and sediments since the late 1970s. She was also, at the time, funded by the National Oceanic and Atmospheric Administration (NOAA) to test for contaminants in the bight. Apprised of the contamination, the federal agency reached out to the California Attorney General. The bight's Channel Islands were a national monument that had once provided habitat for several species that had been disappearing for decades. With evidence of DDT contamination accumulating, the US Department of Justice planned to sue Montrose for destroying natural resources protected by federal law. Because there was contamination in two places, out by the federally protected Channel Islands and close to shore, right off of California's Palos Verdes peninsula, the department invited the state to join the suit.

The suit, *US v. Montrose*, was filed in 1990. By 1993, as Wolff and Toniolo's study was making headlines, it was clear the suit wouldn't be resolved anytime soon. John Saurenman, a lawyer with the Land Law section of the California Attorney General's office, was finding the case to be one of the most intensively litigated he had ever experienced. The plaintiffs needed scientific evidence that indisputably linked DDT in the ocean to wildlife harm. The strongest evidence linked the decline of bald eagles, peregrine falcons, and brown pelicans to DDT-contaminated fish, but the food chains all pointed to fish contaminated by the DDT discharged by the sewer plant, not the DDT dumped in the ocean. There simply was no scientific, and therefore legally tenable, way to tie the ocean-dumped DDT to environmental damage. And the case's success, Saurenman knew, "was all a question of what we could prove."

Amassing the strongest scientific evidence was one thing; dealing with the defendants was another. On top of its relationship with Stauffer Chemical, Montrose had been bought by Baldwin Rubber Company, and the resulting Baldwin-Montrose company had later merged with boat manufacturer Chris-Craft; Stauffer itself had been sold twice in the 1980s alone, its different chemical divisions going to different companies. Before long, *US v. Montrose* had more than half a dozen defendants. The defendants appealed every decision not in their favor. They brought in more than a hundred other defendants, local governments that they argued were just as responsible for the pollution. Scientists for the plaintiffs, meanwhile, were sampling water and sediment, collecting eggshells, and assessing kelp beds and more in attempts to demonstrate the contamination's harms in the midst of litigation. "It was a bad place to do science," Saurenman said.

The plaintiffs settled with the local governments first and then turned back to the corporate defendants. As the case worked its way through the courts, the EPA declared the old Montrose plant and its sewage outfall a Superfund site. Thirty years later, parts of the colossal case would still be pending.

Chapter 25

DISRUPTION

As an academic, Theo Colborn had quickly gotten used to speaking to audiences of students, colleagues, and other people interested in, or invested in, her research. After the Wingspread conference, all of her talks concerned one subject: endocrine disruption. She often focused her talks on pollution and wildlife effects in the Great Lakes because it was the place where her research began and it was what she knew most about. And she took advantage of the chances she was offered to talk about her work, for she wanted to share what she had found and learn whether the hypothesis also resonated with other scientists.

But after the first few Wingspread consensus publications came out, Colborn started to notice something. Every time she was scheduled to present or speak somewhere, someone representing an industry group, such as the American Plastics Council or the American Petroleum Institute, soon got on the agenda as well.

She wasn't surprised that the theory had detractors. A chemical industry toxicologist dismissed it in the *New York Times* as nothing to worry about. A scientist from the ACSH, which was known for its chemical, plastics, and petroleum industry funders, called it "a lot of baloney."

But denouncing endocrine disruption to reporters was one thing. Following Colborn wherever she spoke was another. They were "basically stalking me," she believed.

She gave herself a new agenda. She needed to try to get other scientists to talk about endocrine disruption, especially to journalists and reporters. It wasn't easy, though. "We need scientists who will speak out," she said, "but so few scientists will." So she, Pete Myers, and Dianne Dumanoski wrote as fast as they could, on a set of deadlines that made it feel, Dumanoski said, like they were writing *Fear and Loathing in Las Vegas*.

And then Dumanoski was diagnosed with breast cancer. The three of them agreed to keep it a secret. Like Rachel Carson had so many decades before, they too feared the attacks they would face if anyone found out that Dumanoski was sick.

•

Mary Wolff, meanwhile, was fielding requests left and right. The National Breast Cancer Coalition had reached out to her, and so had scientists across the country. Her lab had found a connection that suddenly everyone wanted to know more about.

She lent her expertise where she sensed it was needed. She participated with a team of colleagues led by an outspoken epidemiologist, Devra Lee Davis, to publish a hypothesis that "xenoestrogens"—their preferred term for synthetic chemicals that mimicked or affected estrogen—were contributing to the rising risk of breast cancer, possibly by interacting with genes that made people susceptible to the disease. (This hypothesis, in fact, was an elaboration of the one Carson had outlined in her letters to Spock decades before.) She collaborated with a different team of researchers, led by Oakland, California, epidemiologist Nancy Krieger, who had more than fifty thousand blood samples taken from white, Black, and Asian women enrolled in the same health care plan back in the 1960s. As Wolff had done with Paolo Toniolo, she and Krieger designed a nested case-control study to illuminate the relationship between breast cancer and organochlorines. Then the epidemiologist carrying out the congressionally funded Long Island study, Marilie Gammon at Columbia University, asked her to analyze samples for that study too. For the next several years, these types of analyses were all anyone seemed to want her to do.

But none of the studies found what she and Toniolo had found. The Oakland study found no difference in DDE levels in women with breast cancer compared to healthy women, unless they were Black. The hard-fought, highly publicized Long Island study was slowly getting underway, collecting enormous quantities of data from the women it enrolled. But so far it, too, was not finding a consistent connection between women's exposure to environmental chemicals and risk of disease. While the Long Island study was starting, the Chemical Manufacturers Association funded a Swedish scientist working at Harvard, Hans-Olov Adami, to review the medical literature on organochlorines and women's estrogen-related cancers to date. Adami and his coauthors noted that chemicals such as DDT and DDE were far less potent than other estrogens to which women were exposed, such as those in birth control pills and hormone-replacement therapy. They looked closely at Wolff's study with Toniolo and the study she worked on with Krieger, which they classified as the two strongest studies out there on the topic. But they concluded that with these conflicting findings, it simply wasn't possible to accept or reject the hypothesis that exposure to such chemicals in the environment could increase a woman's breast cancer risk.

By the time Adami's paper was published in 1995, it was becoming increasingly clear to Wolff that many of the discussions about the connection between breast cancer and xenoestrogens were missing something. The Wingspread scientists had pointed out that endocrine disruptors often seemed to cause problems by affecting the process of development during a specific window of time during that process. In other words, a xenoestrogen might cause harm if a person was exposed at a particular point in development, but it might not cause any harm if they were exposed before or after that time. As it happened, that particular idea was just about to get a lot more attention.

•

Charles Wurster put on a blue suit, a white shirt, and a burgundy tie, clipped with a pin in the shape of an osprey in flight. It was silver, like his hair. He was retiring from Stony Brook, and the Marine Sciences Research Center on campus was holding a symposium to mark the occasion and name the day, April 25, 1995, after him in his honor.

Ian Nisbet came down to talk at the symposium. So did a wildlife biologist from the New York Department of Environmental Conservation, who described the return of Long Island's osprey, endangered in the seventies and now nesting in the hundreds once again. Lorrie Otto flew in from Wisconsin to listen, dressed from head to toe in cherry red. William Ruckelshaus, the EPA administrator who had canceled DDT's registration back in 1972, spoke at the dinner that followed at a nearby country inn that evening. Students sat at one table, scientists and their families and friends at the others. Wurster sat on a dais at the front of the room while Ruckelshaus urged them all to see government as their ally and not the enemy.

For Wurster's organization, the Environmental Defense Fund, had a well-known reputation for suing government by then. Other environmental organizations had made a two-decade habit of doing the same by then, too. In fact, the Natural Resources Defense Council had just successfully sued the EPA to get it to enforce the Delaney Clause, which the agency had all but stopped doing in the 1980s. But the EDF had also grown way beyond its old "sue the bastards" motto. Its staff had come up with novel ways of curbing air pollution and reducing commercial waste. The organization now had 300,000 members and a $25 million budget. On the day of Wurster's symposium, a reporter attended to try to learn what had made the environmental movement so fervent back in the seventies and what kept it going in the nineties.

She also interviewed him at his home, by the floor-to-ceiling windows of his living room, which overlooked a forest that filtered a view of the Smithtown Bay. "Right now," Wurster told her, environmentalists were concerned "with maintenance, with not losing ground." Then he paused. The reporter noticed that the birds outside were quiet. "I guess if I were to involve myself in anything in that same way again, it would be in the area of how pollutants can function as hormones," he said. He recalled a scientific study he had read decades before. Back in the late 1940s, a Syracuse University scientist and his graduate student had injected DDT into forty young leghorn cockerels for several months. The roosters didn't die, and they didn't get sick, but they also didn't look like males. Their testicles were minuscule, just 18 percent of average size, and their combs and wattles were small and pale. The changes made them look like hens.

Now, it seemed to Wurster, studies like that were coming out every day, and they were getting a lot of attention. A University of Florida scientist had found that alligators in Florida's Lake Apopka, contaminated with a DDT-like pesticide back in 1980, had extraordinarily low testosterone levels and unusually small penises. Just recently, University of North Carolina scientists had produced male rats with chest nipples and small reproductive organs by feeding them—and pregnant rats—DDE. More and more scientists all over the US were finding what a California scientist had found back in the 1970s: in coastal bird populations exposed to DDT, the numbers of males plummeted so dramatically that female birds started nesting together, producing sterile eggs. The papers called them gay gulls. Conservative commentators denounced the studies as academic absurdity. But Wurster knew it was all frighteningly true. The California nesting sites were on the Channel Islands, which sat in the Pacific just west of Montrose's DDT plant, which had flushed its waste into Los Angeles County sewers that emptied into the ocean two miles off the coast.

And DDT wasn't even a particularly strong estrogen mimic. It was just the most commonly known of the dozens of chemicals, many of them organochlorines, increasingly coming to light as estrogenic. Some studies were now pointing to their effects on people. European scientists were reporting declining sperm counts in men and reproductive abnormalities in boys. American researchers were also starting to document the abnormalities. Wurster told the reporter that it all amounted to the "feminization of the planet." The pesticide problem had changed so much since his early days.

"You know that old saying," he said. "You can't go home."

•

Rachel Carson had read the same rooster study Wurster had, long before he had come across it. As she had pieced together her chapter on pesticides' effects on cellular oxidation, the process by which cells produce energy, the study had seemed proof of something significant. DDT and other pesticides were "crowbars" that "wrecked the wheels of oxidation," she wrote. When oxidation went awry, cancer, birth defects, and reproductive failure could follow. The last of these effects was probably the result of something well documented in the case of DDT: scientists had repeatedly found it accumulated in the sex organs of birds. They had also found that when birds and other animals were fed DDT, like the cockerels in the Syracuse study, their sex organs atrophied. She believed it was because the DDT concentrated there was destroying their ability to produce energy.

Decades after Wurster saw the study, Colborn had come across it, too. But when she read it, something else about it stood out. The Syracuse scientists had pointed out that DDT was structurally similar to DES, a well-known synthetic estrogen. They had also pointed out that some mammals were "extremely sensitive" to small amounts of DDT. Colborn wondered how so many scientists had missed the study's implications for so long. She wondered in particular how Carson had read the study and interpreted its significance so differently. She wondered if it wasn't because Carson was simply preoccupied with her own debilitating cancer. "Our obsession with cancer blinds us to other dangers," she thought.

To Colborn, Dumanoski, and Myers, the stories of DES and the DDT-exposed roosters, when considered side by side, explained so much. The cancers in the daughters of the women who took DES proved that chemicals crossed the placenta, that amounts too small to affect adults could affect a baby's prenatal development, and that the consequences of such an exposure might not be realized for decades. The rooster study showed that DDT was, like DES, capable of acting like a hormone. It mattered that DES and DDT resembled each other: DDT, like DES, could bind to receptors on the surfaces of cells that were meant for estrogen. It also appeared to block the male hormone androgen from binding to its own receptors. But it mattered even more that neither DDT nor DES resembled estrogen itself. A pregnant woman's blood contains special proteins that soak up estrogen to protect the developing fetus from it—but those proteins didn't see estrogen-mimicking chemicals in the same way. "The world is full of hormone disruptors," wrote Colborn and her coauthors in their book, *Our Stolen Future*. And their ability to act as impostors one moment and not the next meant they were wreaking "hormone havoc."

Our Stolen Future was published in March 1996. The book's cover featured a coiled human fetus, glowing white against a black background. Vice President Al Gore, an ardent champion of environmental causes, had agreed to write a foreword. The book, he wrote, was in many respects a sequel to *Silent Spring*. "*Our Stolen Future* raises questions just as profound as those Carson raised thirty years ago—questions for which we must seek answers," he wrote. The book catapulted Colborn, a year shy of seventy, into celebrity status.

News outlets echoed the vice president, calling the book the next *Silent Spring* and Colborn the next Rachel Carson. Colborn, in good health, was suddenly everywhere: the morning news shows, the evening news, cable news, public television, and radio. Her publisher hired a public relations firm to help her craft a simple, media-friendly message: plastics and pesticides that were ubiquitous in the environment and the food supply were interfering with the chemical messages that tell babies how to develop. Safety testing in the US wasn't protecting children or wildlife.

"The safety net that we used as our model was this 70 kilogram adult male," she said. "We did not look at what happens during embryonic development in the womb or in the egg." It was an argument that built on those made by Morton Biskind in the 1950s, Rachel Carson in the 1960s, and Kathleen Kreiss in the 1980s.

The first print run of *Our Stolen Future* sold out fast. Environmental groups feted Colborn and showered her with awards. The National Academy of Sciences quickly convened a panel to study the issue. Just as quickly, the EPA announced that it would change how chemicals were tested for safety in order to look for reproductive effects in animals tested in the lab.

•

But from the get-go it was hard to catch a news report or read a review of the book without hearing about how controversial it was. Some scientists said it was an important warning, said a reporter on CNN, but others insisted it was a scare tactic. Colborn wasn't surprised to hear her work cast that way, especially after sensing for years that industry representatives were following her everywhere she spoke. Now she learned that she wasn't wrong. An unnamed environmental group gave the *Washington Post* documents showing that the Chemical Manufacturers Association, the Chlorine Chemistry Council, the American Plastics Council, the Society of the Plastics Industry, and the American Crop Protection Association had all attempted to launch a coordinated attack on her book well before it was published. They had also created a new group, the Endocrine Issues Coalition, to fund "sound science" on such topics as the estrogenicity of nonylphenol, the relationship between PCBs and endometrial cancer, and the link between breast cancer and DDT.

Although the industry groups struggled over who should be out in front on the issue, at least one, the Chemical Manufacturers Association, pulled together a twenty-three-page press kit for reporters, summarizing scientific studies that contradicted the book's claims. The group also had help from impartial-sounding allies. Elizabeth Whelan's ACSH pulled together a thirteen-page rebuttal of the book's arguments and sent that to reporters too. When Whelan and her staff took calls from journalists asking for opinions on the book, they told them the whole idea of endocrine disruption was nothing but "innuendo on top of hypothesis on top of theory." A group called the Advancement for Sound Science Coalition (TASSC) also campaigned against the book, sending daily press releases to reporters as the book launched. One release included a list of scientific experts willing to talk to reporters about the book's flaws. The book contained as much science fiction as *Jurassic Park*, said a Dartmouth chemistry professor on the list.

The carefully vetted list of experts was a well-tested strategy that TASSC had put to the test elsewhere. For the coalition was the creation of tobacco giant Philip Morris. The company and its public relations firm, APCO Associates, had created the beneficent-sounding organization three years earlier to fight back against the growing number of laws banning smoking in public. Within no time, TASSC's network of scientists, lawmakers, and politicians was all over network and cable news, radio, and the papers calling scientific studies on secondhand smoke "junk science." The idea of "junk science" was so effective they soon wielded it against studies of global warming and endocrine disruption, too. It was all part of a plan, TASSC told its members, to keep damaging research out of the courtroom and to fight government regulation. These were goals the industry shared with the chemical and fossil fuel industries, too.

Two months later, in early May, *Our Stolen Future* was released in Europe. A man who described himself as the director of the environmental unit at the Institute of Economic Affairs in London, Roger Bate, wrote a review of the book for the *Wall Street Journal Europe*. According to Bate, the claims about endocrine disruption were vastly overstated. To prove it, he cited a report published by something called the European Science and Environment Forum (ESEF), which described the natural hormone mimics found in nature and in such foods as carrots, coffee, potatoes, and wheat. They were all harmless, he wrote. He criticized Colborn and her coauthors for calling for chemical bans on the basis of "thin" evidence. Any further bans, he argued, would cause more problems than they would solve. "It would be self-defeating to push for an outright ban of an estrogen-disrupting pesticide like DDT," he wrote, "when we know for sure that many people are dying from mosquito-transmitted malaria."

The ESEF, it turned out, was Bate's own creation. So was the environmental unit at the Institute of Economic Affairs, which he had founded two years before. Later that summer, in August, a staff member at Bate's institute included the book review in a package of materials addressed to the head of corporate communications at British American Tobacco. Bate wanted the company to know about his unit and its core mission: to explain how "private property, a dynamic market economy and the rule of law protect and improve the quality of the environment." The tobacco industry, he felt sure, would support him in this quest.

He was right. His bit of self-promotion would put the tobacco industry behind an emerging call to bring back DDT—not because the industry wanted to use it as a pesticide but because the industry wanted to use it as a symbol of government regulation gone horribly wrong.

Chapter 26

DELANEY FALLS

Eligio "Kiko" de la Garza II liked to disarm his audiences with a couple of self-deprecating jokes before getting down to business. He was first elected to Congress back in 1964, when his election-day win made him the first Mexican American to represent Texas's southernmost district, bordering Mexico. By the time Bill Clinton became president in 1992, de la Garza had been in Congress for nearly thirty years and had been chair of the House Committee on Agriculture for more than ten. He had championed farm bills, drought relief acts, aid to farmers facing losses, and programs to expand the global market for US agricultural products. He was a Democrat with a conservative reputation, and when Clinton was elected, he had invited de la Garza to get involved in an administration plan to reorganize the federal government.

But de la Garza, stout with a square jowl and a few wisps across the top of his head by now, had always put agricultural causes first. In 1994 he tried to pass a bill that, he said, would solve a "major crisis" in agriculture: the need for so-called minor pesticides. The crisis was brought to his attention by an alliance of farmers who grew "minor" crops, which included vegetables, fruits, seeds, and nuts. A 1988 amendment to the Federal Insecticide, Fungicide and Rodenticide Act (FIFRA) required chemical companies to reregister older pesticides to ensure that they met modern safety standards.

The process of testing a chemical was expensive, so pesticide makers were focusing on the chemicals used on major crops, such as wheat and corn. Minor crop farmers said the tightened regulation meant that the pesticides they used were likely to disappear. They wanted an expedited review process so that their preferred chemicals could be applied to crops without going through the formal registration process required by law. With House Bill 967, the Minor Crop Protection Act of 1994, de la Garza obliged.

When he presented the bill on the House floor in early October 1994, he told his congressional colleagues that his committee had originally hoped to introduce a rewrite of both FIFRA and the pesticide-related areas of the Federal Food, Drug, and Cosmetic Act. But because those plans were stalled, he was instead presenting a plan to give chemical manufacturers more time to use existing safety data on pesticides, more time to submit new data on pesticides, and waivers of certain safety data requirements, all "to provide farmers and public health agencies with continued access to these important tools."

Representative Pat Roberts, a Republican from Kansas with just as much dedication to agricultural concerns but far fewer years in Congress under his belt, chimed in to support de la Garza's bill. He also raised the specter of rising fruit and vegetable costs for consumers if pesticide manufacturers had to pay the high cost of keeping minor pesticides on the market. "There are many other pesticide issues outstanding, including modernizing the Delaney clause," he said, invoking the 1958 amendment to the Food, Drug, and Cosmetic Act that banned all chemicals suspected of causing cancer in humans from the food supply. "But for today, I urge my colleagues to support H.R. 967."

Then Henry Waxman stood up. The mustachioed California Democrat, chair of the Energy and Commerce Subcommittee on Health and the Environment, had spent the year grilling tobacco executives, getting them to swear under oath that nicotine was not addictive, in hearings that later revealed that the companies had long known that nicotine *was* addictive and that cigarettes caused cancer and heart disease. The Minor Crop Protection Act, Waxman said, was not what it seemed. It was controversial, he said. It had been rushed out of committee. And if enacted it would authorize the use of "potentially dangerous pesticides" that could harm the public, especially children.

Waxman's objections made little difference. In a vote the next day the proposed amendment passed. But in the Senate, lacking support, it never even made it to a vote. A month later, on Election Day, the Republicans won control of both houses of Congress for the first time since the 1950s. The House elected Georgia Republican Newt Gingrich Speaker. Gingrich and his fellow Republican leaders' legislative agenda, which they dubbed their Contract with America, prioritized cutting the federal budget and reining in federal regulatory authority, especially when it came to protecting the environment. Elizabeth Whelan and her ACSH cheered them on.

The "'Contract with America' calls for the rejection of junk science and a return to 'mainstream' views," she wrote in one of her many op-eds. "A disembowelment of Delaney would be an excellent start."

•

Whelan had been on Larry King's cable news show before, after the Alar apple scare of 1989. King had introduced her then as a "critic" of environmentalists, who were enraged when tests by the Natural Resources Defense Council had found residues of Alar, a growth-promoting chemical, in apples and apple products. Alar had no place in a food so widely consumed by children, environmental groups said, because it caused cancer in lab rodents. Under the Delaney Clause, that was illegal. Whelan had a different view. "If we ban everything that causes cancer in mice, we're going to have very little to eat," she told King.

King took some flak for not mentioning that Whelan's organization was funded by industries, including the chemical and food industries, and a producer for his show publicly apologized. But King wasn't alone: dozens of newspapers, wire services, and television news shows had been featuring Whelan's opinions on chemicals in the food supply without disclosing her industry support.

As de la Garza's bill was discussed on the House floor in October 1994, King invited Whelan back, this time to weigh in on a new NRDC report; this one claimed that children were routinely exposed to carcinogens in the environment. "Larry," she said, "there is no mainstream scientific evidence that environmental chemicals, whether it's pesticides or formaldehyde or asbestos in school, play any role whatsoever in the causation of cancer."

"Are you saying that no pesticides, no chemicals cause cancer?" King asked.

"I'm not saying that," she said. "It's the dose that makes the poison."

King turned to Lawrie Mott, the California scientist who had authored the NRDC's report. The argument about dose, Mott said, didn't take children's intake levels into account: "It's a simple biological fact that children eat more food, breathe more air, and drink more water, as a percentage of their body weight, than adults do." The government, she said, needed to strengthen its pesticide safety standards to protect children.

Whelan jumped in. The vast majority of cancer was caused by tobacco, she said. Sunlight, drugs, and workplace chemicals caused much of the remaining cases, she said. And the high doses of chemicals causing cancer in mice said nothing about the real dangers to children's health. Then Mott cut in over her. Letting people get cancer before declaring something carcinogenic, she said, was "morally reprehensible." Then Whelan cut in again, and King cut them both off. "One at a time, one at a time," he said.

Whelan was well practiced at staying on point, and there was one particular point she wanted to be sure to make. She had been targeting the Delaney Clause ever since she founded her organization, back in 1978. Repealing the clause was a core part of her organization's mission. She had been working for years to convince top scientists in government and research organizations, such as the surgeon general and the National Cancer Institute, to speak out about the problems with what she called Delaneyism. But she also needed to convince the public. She went straight for one of her favorite autumn arguments. "If we were to err on the side of safety, the Food and Drug Administration would ban Thanksgiving Dinner" because of the natural carcinogens found in every course, from the turkey to the pumpkin pie. "I get very concerned when I see the EPA running after banning pesticides and other useful chemicals," she said, "without science to back them up."

"That's wrong," said Mott.

"It's toxicologically correct," said Whelan.

"I think the public should know where the majority of your funding comes from," said Mott. "The majority of your funding comes from the oil industry, the chemical industry, and the food industry—" Whelan cut in. Mott's funding came from a family who believed environmental chemicals had caused their child's cancer, she said. Mott insisted the family had no editorial input. Whelan said the same was true for her funders and tried to divert.

"I think we should drop that subject," she said, "and get to the science."

King interrupted them both to cut to a break. "Don't go away," he said to the camera.

•

By the mid-1990s, pressure for a major change in pesticide regulation had been building on all sides. The DDT ban in 1972 had heralded a decade of bans: other organochlorine pesticides followed, as did some food additives, such as the dye Red No. 2, and certain drugs, such as DES. The food and chemical industries lobbied against the approach, which the Delaney Clause propped up. Whelan often sent her anti-Delaney op-eds straight to her industry funders because she knew it was just the type of message they wanted to see repeated.

But environmental groups such as the NRDC weren't happy with pesticide regulation either, for the reasons Mott shared on King's program. Neither the Federal Food, Drug, and Cosmetic Act nor FIFRA was written with children's unique vulnerability to chemical pesticides in mind. And the Delaney Clause's focus on carcinogens was far too narrow. New studies were suggesting that pesticides could harm health, especially children's health, in other ways, causing potential harm to the endocrine system, immune system, and nervous system.

The Clinton administration's EPA was also jockeying for reform. EPA administrator Carol Browner noted that despite bans on persistent pesticides and laws meant to keep dangerous pesticides off the market, overall pesticide use had nearly doubled over the previous three decades. More and more health experts were arguing that pesticides put children's health at unique risk. And children's health and welfare were top items on the administration's domestic policy agenda.

Browner was also still reckoning with a 1992 Ninth Circuit Court of Appeals decision in the case brought against the agency by the NRDC, the state of California, and farmworker groups over the Delaney Clause. The plaintiffs had charged that the EPA had broken the law when it had stopped enforcing the Delaney Clause in the 1980s; the agency, in light of a new federal report on the standard's shortcomings, had started to adopt a standard known as "negligible risk" in its place. The concept, advanced by some epidemiologists and toxicologists in the seventies, held that a chemical was safe enough for consumption if it posed a risk that was negligible: no more than a single additional case of cancer for every million people consuming it over a hypothetical seventy-year life span. The law, argued the plaintiffs, said that when it came to carcinogens, the risk should be zero.

The defense, meanwhile, pointed out that there was another law that applied. FIFRA, enforced by the EPA, had been amended to allow trace amounts of pesticides in food, including animal carcinogens, as long as they posed no "unreasonable risk" to consumers. The law applied to raw crops, whereas Delaney still applied a zero-tolerance approach to processed foods. "This basic disagreement between the two statutes has tied the EPA in knots," reported the *New York Times*.

In October 1994, the same month that de la Garza presented his bill and Whelan and Mott debated on Larry King, Browner's EPA settled the suit. Under the settlement the agency agreed to take action against three dozen pesticides over the next two years, which meant reviewing data on three dozen potentially carcinogenic pesticides to determine whether Delaney applied. The agency also agreed to review the data on eighty-five other pesticides to determine if their residues were found in processed foods. These actions, newspapers reported, would probably lead to dozens of pesticide bans over the next few years.

Browner, who had taken office at age thirty-seven, with two years as the head of the Florida Department of Environmental Regulation under her belt, appealed to Congress for help. "We do not believe that consumption of these pesticides as residues in processed food products is a threat," she said. "These foods are safe." With analytical chemistry now able to detect parts per *trillion* of pesticide residues in foods, a negligible risk standard would be more realistic. The Delaney Clause, she said, was out of date: "The entire food safety program requires overhaul."

Browner's claim that foods were safe irked environmental groups. But Whelan quickly came to her defense. "Browner and the EPA need our support and encouragement," Whelan wrote in an editorial that ran in papers across the country. "It takes a strong determination and a commitment to do what is scientifically correct, not politically correct." She appealed to funders for added support. ACSH was leading the battle to defend "safe use of pesticides and chemicals," she told them, and to decry "the numerous unnecessary government bans, interventions and regulations that stifle America's free enterprise system."

Corporations responded. In the year that followed, Exxon, Dow Chemical, Ciba-Geigy, American Cyanamid, DuPont, Union Carbide, Uniroyal, Coca-Cola, Eli Lilly, Shell Oil, Georgia-Pacific, Johnson & Johnson, and others donated to support what Whelan called "the unique voice of reason" in national debates over the environment, human health, and disease.

•

With the Republicans in the majority, de la Garza stepped down as chair of the House Agricultural Committee, and Pat Roberts stepped into the role. Clinton asked Congress to find a way to reform pesticide regulation, and an unlikely duo, Waxman and Virginia Republican Thomas J. Bliley Jr., chair of the House Commerce Committee, ultimately teamed up to help make it happen. Bliley, a onetime undertaker and the former mayor of Richmond, was a tobacco industry loyalist. He had a neat sweep of white hair across his brow and a framed picture of every Philip Morris brand of cigarette in his Washington office. Waxman, whom Bliley liked to refer to as the gentleman from Hollywood, was one of tobacco's fiercest opponents. For different reasons, they both wanted to see Delaney fall.

Waxman believed that Congress should enact safety protections that took more than cancer into account and that Delaney could be replaced with a single standard that applied to raw and processed foods. The laws governing raw foods, he said, essentially let farmers use "chemical weapons to treat their crops." As chair of the Commerce Committee, Bliley had vowed to secure lighter regulations of businesses across the board, and a relaxation of the Delaney Clause at the very least. He also knew that without bipartisan support, President Clinton would veto any bill that wholly catered to the food industry. Coming from opposite directions, he and Waxman had both arrived at the same place.

By the summer of 1995, Congress was considering several food safety bills, most of them with the Delaney Clause in their sights. One, drafted by Bliley and his committee's senior Democrat, John Dingell of Michigan, proposed fixing Delaney and including stronger protections for children. Another, a bill to stall regulation by imposing cost-benefit analyses, drafted by the food industry and sponsored by Kansas Republican Bob Dole in the Senate, proposed that no federal agency could ban a product that posed "negligible or insignificant human risk." It didn't mention the Delaney Clause, but it aimed straight at it.

None of those bills made it very far. But the administration was intent on resolving what many had taken to calling "the Delaney paradox," from a recent National Academy of Sciences report on the clause. When a draft of Bliley and Dingell's bill landed on the desk of an EPA appointee named Jim Aidala, he realized that it had all the elements of a potentially viable bipartisan compromise.

•

The EPA had been a target of political conservatives since it was created. And from that same moment, its pesticide program faced a workload it could never manage, for it was tasked with reviewing decades of data on thousands of chemicals to ensure they were in compliance with ever-changing federal regulations. Right after the EPA was created, in fact, Congress passed amendments to FIFRA that required manufacturers to submit additional health and safety data in order to register a pesticide. A few years later, an investigation revealed that hundreds of pesticides had been registered with safety testing results supplied by a firm with a reputation for falsifying data. The finding led to yet another category of registration, conditional registration, with its own data-reporting requirements. From the beginning, the EPA's critics took advantage of the agency's "crushing backlog" of pesticide data to claim that the pesticide program accomplished so little it might as well be replaced by a bunch of chimps. But Aidala knew that wasn't true.

As a graduate student in sociology at Harvard in the seventies, Aidala had studied the EPA's pesticide program and interviewed its staff as a case study in understanding organizational behavior. The work made him an expert in the program's function. In 1979 the EPA hired him as a program analyst in the agency's Pesticide and Chemistry Office. He lost the job under the Reagan administration, the funds supporting his position eliminated overnight. But he eventually landed a job as a subject expert on pesticide regulation at the Congressional Research Service, which everyone on the inside thought of as the shadow government. "It was the thing that made it look like staffers knew something about anything," he said.

When Clinton was elected, Aidala went back to the EPA as a senior-level political appointee in the agency's Office of Prevention, Pesticides and Toxic Substances, which is where he was when members of Congress began floating food safety bills to address the many problems with pesticide regulations. An administrator above him, Lynn Goldman, had pulled him onto a team of staffers tasked with figuring out all the details of what such a bill should look like. Now he was working like mad, still behind the scenes, to iron out a compromise solution that would make both environmentalists and conservatives happy. Behind-the-scenes folks like him, he said, called the two constituencies the pinheads and the coneheads, after a *Saturday Night Live* sketch about space aliens. Neither of them seemed fully of this world.

A few things were clear. An EPA conference had determined that the endocrine-disruption hypothesis was valid and that the area deserved more attention. A pair of National Academy of Sciences reports had concluded that pesticides did pose an added risk to infants and children and that Delaney's narrow focus on carcinogenicity might actually be keeping safer chemicals off the market. A pesticide could be a potent neurotoxin and still be registered for use, for instance, whereas a pesticide with a minuscule cancer risk could not be. Delaney's defenders liked the fact that the clause was so cautious, but it was just as easy to argue, by 1995, that its omissions made it risky, especially because, as organochlorines had disappeared from the market, the share of organophosphates in use skyrocketed, even though they were notoriously and acutely neurotoxic.

The team Aidala sat on—with representatives from the USDA, FDA, and EPA and staffers for Bliley, Dingell, and Waxman—started scheduling meetings. They met with pesticide makers. They sat down with big environmental groups, such as the EDF and the NRDC. They met with trade association representatives, congressional influencers, and top White House staff. Eventually, they had a roughly hundred-page bill that everyone—pinhead or conehead—seemed happy with.

•

On July 23, 1996, Roberts presented H.R. 1627, the Food Quality Protection Act of 1995, for a vote in the House. As he did so, he acknowledged de la Garza as the "godfather" of an effort to modernize pesticide regulation that was guided by "science, not emotion."

De la Garza thanked Roberts in return. The biggest hurdle in the whole process, de la Garza said, had been the Delaney Clause, a "political outgrowth" of the public's fear of cancer back in the 1950s. Bliley put a finer point on it: the year Delaney was enacted, he said, "was the year Fidel Castro came to power in Cuba. Like Castro, the Delaney clause has cast a long and dark shadow over the years." Bliley went on. Delaney had held the nation back, he said, because it had "frozen science for 40 years."

Waxman followed. He wanted his House colleagues to know that despite differences of opinion and different starting points, they had arrived at a bill they all agreed on. "We have a good bill. It is a good compromise," he said. "The American people should look upon this with favor."

Aidala, for his part, was proud of the bill. It was tough, in his view, and modern. It met everyone's requirements. And that hadn't been easy. Partisan politics were so rancorous that a budget dispute had shut down the entire federal government late in 1995. But the coalition behind the bill had succeeded nonetheless. It had produced a bill that the chemical industry supported, as did the food industry, agricultural groups, public health groups, and environmental groups.

The compromise bill did away with the Delaney Clause, but it required the EPA to consider other health effects of pesticides when reviewing safety data, including endocrine, immune-system, and nervous-system effects. It also required the agency to consider pesticides' effects on infants and children by instituting a "ten-fold" safety factor. If studies suggested that 1 ppm of a chemical was safe, then the safe dose for children would be set at 0.1 ppm. The law required the agency to consider the safety of aggregate effects of pesticide exposures and also the effects of extreme diets—like people who ate copious apples (such as the Montana girl killed by the arsenic-laden fruit in the 1930s) or only meat. And in place of Delaney's no-tolerance standard, it was all guided by the concept of negligible risk.

The bill passed the House that day and the Senate the next. President Bill Clinton signed it into law on August 3. "If Waxman and Bliley are together on this," the president's chief of staff reportedly said, "we're for it." And with that, the Delaney Clause was dead. Its mourners, few in number, called it the toughest public health law ever. Its detractors said it was way too out of date to mourn. The *Wall Street Journal*'s editors later gave Whelan credit for taking on the clause and staying the course for so long. But they sounded almost sad to see it go. The government's "outrageous interpretations of the Delaney Clause," they wrote, "was one of our favorite fights."

Satisfied, Whelan moved on. She was done with Delaney, but she wasn't done with many other issues, including DDT. She set her staff to updating the ACSH *Facts Versus Fears* report, a summary of the "greatest unfounded health scares of recent times." DDT, as ever, topped the list. The report cited the studies that found no connection between DDT and cancer. It tallied the millions of malaria deaths in countries that had abandoned it after it was banned in the West. It acknowledged that populations of birds, especially falcons and eagles, had rebounded after the ban. But it denied that this had anything to do with DDT.

Chapter 27

BRING BACK DDT

The crows died first. Hundreds of them, all across New York City, followed by three flamingos, a cormorant, and a bald eagle at the Bronx Zoo. People fell sick next, with fevers, brain inflammation, and muscle weakness so severe it was paralyzing. Dozens were hospitalized. The most elderly died. Tests at the CDC and a specialized lab in California soon confirmed the cause: West Nile virus, a mosquito-borne killer from Uganda never before seen in the United States.

Before the test results were in, Mayor Rudy Giuliani made a quick decision. Emergency-management officials would spray the city with pesticides, from Brooklyn to the Bronx and Staten Island to Queens. "The more dead mosquitoes," he said, "the better."

It was the summer of 1999, a season of drought, record heat, and massive blackouts in New York. As Labor Day neared, helicopters billowed white clouds of malathion over northern Queens, the South Bronx, and Central Park. Where tall buildings got in the way, convoys of pickup trucks rolled through, fogging the streets with pyrethrin. The outbreak—and the cure—were "like something out of the middle ages or the backwater of a third world country," said news anchor Tom Brokaw on *NBC Nightly News*. New Yorkers cleared the streets; those who could fled the city entirely. But while some were escaping the virus, others were worried about the spray.

A man in Harlem said he didn't have time to flee a truck's flume. A woman in Whitestone said she was grilling when a helicopter drenched her. One particularly pissed-off group of New Yorkers secured a lawyer, dubbed themselves the No Spray Coalition, and filed an injunction to halt the spraying. "Almost every product brought to us by the highly profitable chemical industry is said to be harmless until, the hard way, we frequently find out otherwise," they argued. "During the 1950s, even deadly DDT was said to be harmless!" The judge heard their arguments and then sided with the city—which continued to spray.

As it did, experts and negotiators convened by the United Nations Environment Program met in Vienna and then again in Geneva, a year into the process of drafting an international treaty to protect the global environment from so-called persistent organic pollutants. The POPs, as they were called, were chemicals that didn't break down in the environment, accumulated in individual organisms, and concentrated in organisms at the top of the food chain. There were twelve of them, and environmental groups and journalists had taken to calling them the dirty dozen. Nine were pesticides, including DDT.

DDT, still manufactured in Asia and used in Asia and Africa, had the international group of negotiators sharply divided. It was unique among the POPs. It shared all of their environmentally damaging characteristics but alone possessed a clear and unmatched benefit to human health. Some believed the pesticide belonged, as a matter of course, with the other POPs, on a list of those that would soon no longer be used anywhere in the world. But others wanted the treaty to include a DDT exception so that the pesticide could be used to control malaria and other mosquito-borne diseases in the world's poorest nations, at least for several more years; at the time, twenty-three countries still relied on it. Taking opposing sides, environmental groups and public health groups both lobbied the negotiators.

"This poses an unusual dilemma," a US state department official involved in the negotiations diplomatically said. Any treaty that included DDT would seem to call for a trade-off between protecting health or the environment—it didn't seem possible to protect both.

At the end of the Geneva meeting, in September 1999, the negotiators adjourned in tentative agreement to grant DDT special treatment. They'd revisit the issue at follow-up meetings scheduled to take place the following year, and they hoped to have something that all parties, representing more than 120 countries, would sign by the end of that year.

In New York, meanwhile, summer's heat eventually gave way to winter's chill, and the city's mosquitoes disappeared. The West Nile outbreak subsided. But in the city's vast subterranean sewer system, mosquitoes survived, harboring the virus within. As the winter thawed and warm weather returned in the spring of 2000, mosquitoes carried the virus back to the city's parks, playgrounds, and streets. Gradually, new infections began to appear in birds and also in people. This time, the disease spread from New York to New Jersey and Connecticut. Again, the city sprayed. West Nile, it appeared, wasn't going away.

•

Alaska had its share of mosquitoes, but in the year 2000, West Nile virus was a remote concern. Far more pressing for Shawna Larson, a twenty-six-year-old with a new baby and a new job, was all of the research coming out showing the prevalence of persistent pollutants in the Arctic. Larson's new job was with an organization known as ACAT, Alaska Community Action on Toxics, which was just finishing up work on a report it had produced with the Department of the Interior. Research over the previous decade, accelerated by an international agreement among Arctic states and Indigenous peoples signed in 1991, had proved something that scientists had first pointed out back in the 1960s. Persistent pollutants such as DDT were finding their way to the Arctic—even if they had never been used there.

The presence of DDT at the Earth's poles had long been a mystery. Rachel Carson had written about the pesticide's presence in Alaskan owls; scientists at the time assumed that they acquired it during migration. Scientists then found the pesticide in penguins and seals at the other end of the Earth, and migration didn't explain it. When, in the seventies, researchers found DDT and other persistent chemicals in the bodies of Alaskan children, they concluded that processed foods must have been the source. But in the eighties and nineties, scientists found chlorinated hydrocarbons, DDT among them, in fur seals, ringed seals, hooded seals, bearded seals, walruses, belugas, porpoises, polar bears, and other species across the Arctic. Gradually, they pieced together what was happening. Large-scale air and water currents were carrying the chemicals around the globe and often northward, where they settled out into the land, the sea ice, and the Arctic waters. In the frigid conditions the chemicals persisted longer than usual because cold temperatures slowed their already sluggish degradation. The Arctic, the Interior Department concluded, had become a global "sink" for persistent pollutants.

Studies showing DDT and PCBs in so many Arctic species inspired an epidemiologist at the National Cancer Institute to ask what this meant for Alaska Natives whose diets included hunted seals and fish. Working with researchers from the CDC and Alaska's Native Health Board, she analyzed serum samples taken from women during a statewide hepatitis B survey in the eighties. Nearly all of them contained PCBs and the DDT breakdown product DDE. Because of global air and water currents and subsistence diets, the team concluded, Alaska Natives were still being exposed to chemicals long banned in the United States.

For Larson, whose father was Athabaskan and whose mother was Supiaq, the research findings were personal. "I don't think we should leave this for future generations," she said. When she joined ACAT, the organization was already on a mission to reduce the flow of pollutants from south to north, in part by lobbying US delegates to take a strong stand in meetings on the evolving POPs treaty. One of Larson's first assignments was to collect signed resolutions from tribal leaders asking the US government to ensure that the treaty embodied the precautionary principle, the idea that a chemical should not be used until it is confirmed to be safe.

By the fall of 2000, Larson had collected a stack of resolutions from more than fifty Alaska Native and American Indian tribes. Her supervisor decided there was no one better to present them at the next POPs meeting, scheduled for early December in South Africa. Larson, who had never left Alaska, boarded a plane with a small delegation from the state and took off for Johannesburg.

•

Amir Attaran had his own strong feelings about DDT's fate under the POPs convention. Attaran, whose parents had immigrated from Iran, was a polymath with a PhD in biology and a law degree. While working on his PhD at the University of Cambridge, he had gotten to know the malaria scientists who worked down the hall. While in law school, he had attended a talk by Ralph Nader, in which the veteran activist bemoaned the lack of global aid for malaria. Attaran went up to Nader afterward to ask a question, and the conversation ended with Nader offering Attaran office space in Washington, DC, to lobby for increased malaria funding. Attaran called the organization The Malaria Project—although he believed that *organization* was too "grandiose" a word for what it was: him, in an office, flying back and forth to Canada as he finished his law degree.

After law school, Attaran accepted a position with the Sierra Legal Defense Fund, just as the organization was invited to provide legal guidance on the POPs convention. But the role made Attaran uncomfortable. Environmental organizations were lobbying for a total DDT ban, but he knew the chemical had a place in malaria prevention. He soon found himself soul searching. On the one hand, he wanted to support his environmentalist colleagues, but on the other hand, he found their position on DDT "just wrong." He talked to his old Cambridge colleagues and his girlfriend, who also studied malaria. He reflected on the defense fund's success in using a letter-writing campaign to lobby Canadian lawmakers to pass an endangered species law. And then he made up his mind. He reached out to one of the malaria scientists he had met in Washington, Mary Galinski, an Emory University professor who had created her own malaria advocacy group, the Malaria Foundation International (MFI). Together, they drafted a letter.

"Dear Diplomat," it began, "we are a group of scientists and doctors who are writing you on account of your participation in the

291

ongoing treaty negotiations. . . . Although we entirely agree that DDT should one day be eliminated because of its known environmental effects, we also believe that human life must not be endangered in reaching that goal."

It was a thoughtful and carefully crafted letter. Malaria was on the rise, particularly in Africa, it said. DDT certainly harmed wildlife but only when overused. The letter explained that there was a way to use DDT that produced only benefits and almost no harm: indoor residual spraying, the very method used across the southern US at the tail end of World War II. DDT wasn't the only tool for controlling malaria, but it was a critical one, alone in its affordability and effectiveness when used indoors. The letter asked negotiators to ban DDT's agricultural uses immediately but allow its continued use in public health until research funded by Western nations produced equally cheap and effective alternatives.

Attaran and Galinski shared the letter with other malaria scientists, who refined its points. One of Galinski's colleagues translated it into Spanish. Someone Attaran knew translated it into French. Then Attaran and Galinski cosigned it and circulated it among their networks of malaria and public health experts, who shared it with their own networks.

By the time of the POPs negotiators' September 1999 meeting in Geneva, Attaran and Galinski's letter bore the signatures of nearly four hundred doctors, scientists, and health economists around the world, three of them Nobel Prize winners. Galinski put the letter and its long list of signatories on MFI's website, highlighting the names of the Nobel laureates in bold.

Then he came up with another idea. He shared the letter with two reporters, one at the *New York Times* and one at the *Guardian*, and offered them an exclusive on the story. Both journalists bit, intrigued by the letter's element of surprise. DDT was a notorious offender, but here were not just a few but hundreds of eminent scientists calling for a reassessment. "DDT, Target of Global Ban, Finds Defenders in Experts on Malaria," reported the *New York Times*. "Millions More Malaria Deaths Feared if DDT Is Banned," reported the *Guardian* in the UK.

Attaran, emboldened by his and Galinski's success, began to send his own essays and op-eds to top medical journals and newspapers. "Imagine seven Boeing 747s, loaded with kids, crashing every day," he wrote. That was how many African children stood to die of malaria each year, he said. DDT remained one of the best ways to reduce the growing number of malaria deaths in Africa, and the argument that it needed to be banned outright to protect people and the planet, he wrote, was "stunningly naïve."

The debate grew. No scientists with some stake in it could sit it out. At conferences, on the pages of medical journals, and in the press, environmental scientists battled against public health experts without end. Walter Rogan, whose research had informed Mary Wolff's breast cancer study, argued that DDT's defenders weren't taking its health risks seriously enough. Breast-feeding was lifesaving, particularly in malarial areas, and DDT exposure curtailed it. Rich Liroff, who had hired Theo Colborn to research Great Lakes pollution, said that even when used sparingly indoors, DDT still ended up in cows' milk and breast milk at levels that exceeded WHO limits. Gilbert Ross, the new medical director of the ACSH (a physician who, incidentally, had been barred from practicing medicine), said the issue was just one more way in which "impoverished humans" were being "sacrificed on the altar of the precautionary principle." The editors of the leading medical journal *Lancet* seemed to agree. The case of DDT, they said, showed that sometimes the precautionary principle needed to be used with caution.

Attaran felt as though the tide had turned. The notion that DDT would be banned like the other POPs no longer seemed a given. To prepare for the Johannesburg meeting, Attaran shifted course. He had taken a new position, at Harvard University, and he now focused on figuring out how to fund poorer countries as they phased out their use of persistent chemicals.

But at the Johannesburg meeting itself, the specific issue of DDT and the larger issue of the precautionary principle still had treaty negotiators stuck in place. The conference was scheduled to adjourn on the evening of December 9, but at the end of the day the negotiators were far from done. They stayed at the table, working through the night. At half past seven the following morning, they had agreed on treaty language. Eleven of the original dirty dozen would be immediately prohibited or gradually eliminated, but DDT would be on a list of its own, its use allowed to control not just malaria but also any vector-borne disease. If countries used it, they needed to notify the World Health Organization. If wealthier nations decided at some point that poorer ones should use a DDT alternative, they would pay the difference in cost. Attaran was overjoyed. "OUR CAMPAIGN TO PREVENT A BAN OF DDT FOR MALARIA CONTROL HAS BEEN SUCCESSFUL!" he posted to MFI's website. "Thanks to all!"

But two more steps remained before the treaty took effect. Nations needed to sign it, and then they needed to ratify it. A signing conference was scheduled to take place in Stockholm the following spring, at the end of May 2001. Once fifty nations ratified the signed treaty, it would take effect. In the meantime, Attaran's letter landed on the desk of a British American Tobacco company executive who found it germane to one of the company's special projects.

•

Bird lovers and ornithologists had their own reason to celebrate at the dawn of the new millennium. Across the US, birds were back. Osprey were building their teetering twiggy nests all across Long Island. Pelican populations had rebounded in Florida. The peregrine falcon had made such a remarkable return that it was removed from the endangered species list, and wildlife experts predicted that the bald eagle would soon follow.

At the same time, however, other birds were dying: crows, least bitterns, black-capped chickadees, mourning doves, mallard ducks, Canada geese, broad-winged hawks, great blue herons, and more. When West Nile virus returned once again to New York in the spring of 2001, it spread to ten states, killing thousands of birds and infecting dozens of people. When human cases appeared in Texas, they triggered panic and pesticide spraying. A writer old enough to remember said it all reminded him of 1949, when polio had devastated the city of San Angelo and the desperate city had saturated itself with DDT.

"A bad dream was back," he said.

In no time at all, his words seemed prophetic, not because West Nile virus shut down cities as polio once had but because, all of a sudden, people all across the country began calling for the return of DDT.

It's time to bring back DDT, said a columnist in Washington, DC. Crank up production, if not for the people of New York, at least for the innocent children of the Third World, wrote a Colorado journalist. Thanks to DDT's ban, the environmental movement bore the blame not just for West Nile Virus but for millions of children dead from malaria, wrote Roger Bate in the *Los Angeles Times*, identifying himself as a scholar on leave from Cambridge. In local papers from California to North Carolina, a former FDA official pointed out the irony that DDT was banned largely for toxicity to birds and now couldn't be used to combat a mosquito-borne virus that was killing birds by the hundreds of thousands. Rachel Carson's legacy, he wrote, was "lamentable."

Franz Froelicher also picked up his pen that spring. He had long since left his parents' home in Ridgewood, New Jersey, and settled in Georgia, following years in Long Island, Alaska, and Mississippi. He was a geologist by training, but by the nineties he was working as a chemist for the US Army Corps of Engineers' Hazardous, Toxic and Radioactive Waste Division in Savannah. His job involved cleaning up polluted sites on government land throughout the Southeast. Sometimes the job involved carting off, burying, filtering, or dredging long-banned DDT. Then West Nile Virus appeared in his backyard. It killed a woman in downtown Atlanta and was killing birds all over the state. And suddenly people were talking about bringing DDT back.

Froelicher had never lost the memory of snacking on the powdery white, slightly sweet crystals in his father's study. His career had brought him back in touch with the pesticide of his childhood. But he was increasingly spending his days of cleanup frustrated and angry. He explained why in a letter to the *New Yorker*, which had just published an article—"The Mosquito Killer: When DDT Was a Gift from God"—that evoked his youth and made him question the present. "Why spend my working hours and the government's money this way," he wrote, "if I do not believe that DDT is harmful?"

Froelicher had nothing to disclose beyond an intimate childhood connection to the chemical. However, most of that season's op-ed writers did. The Washington writer was the executive director of TASSC, the "sound science" front group created by Philip Morris and its PR firm several years earlier. The former FDA official was a TASSC partner. The Colorado journalist was executive director of a "journalism center" directly funded by Philip Morris. Bate, the "Cambridge scholar," had founded the ESEF, which was funded by Philip Morris and British American Tobacco, and whose founding description exactly matched that of TASSC.

In short, DDT's defenders were all part of a campaign dreamed up by Bate, financed by tobacco companies, and designed to protect the global market for tobacco and cigarettes. Bate was a neoliberal think-tank leader with one objective: to advocate for free markets. Meanwhile, the tobacco industry had a separate but related objective: to protect the cigarette market from encroaching regulation. They financed Bate's operation because they were convinced it would serve their own. And DDT had a curious part to play in it all.

For ever since the tobacco litigation settlements that followed Representative Henry Waxman's hearings, the threat of national and international regulation loomed over the industry, and not just in the United States. While the POPs meetings took place, the World Health Organization had developed a Framework Convention on Tobacco Control, which, if adopted, would be the world's first global public health treaty and the first to target tobacco. At the same time, the WHO launched an initiative to consider bringing tobacco under the purview of international drug control treaties, which would treat their product like a narcotic. Both moves had the industry acting fast to keep such added regulatory restrictions at bay.

The British American Tobacco executive who held on to At-taran's letter found it helpful for the company's new corporate social-responsibility objectives and one of its special projects. British American Tobacco had just joined with Philip Morris Tobacco Company and Japan Tobacco International—which together controlled more than 40 percent of the world tobacco market—to launch something called Project Cerberus, named for the three-headed dog that guarded the gates to the underworld in Greek mythology. Project Cerberus had one key aim: to defend the tobacco industry's ability to regulate itself. On their own, Philip Morris executives were also talking about how to defend against regulation. The outcome was a plan for TASSC to engage in an "aggressive year" of activities to promote "science based on sound principles—not on emotions or beliefs considered by some as 'politically correct.'" With the DDT issue heating up, the front group added the controversy to its list of scientific stories to distort. The DDT story served the industry in two ways: it focused global attention on malaria as a health threat bigger than tobacco use. And it implied an inherent hypocrisy in global health efforts led by Western interests, casting doubt on Western nations' ability to set global health agendas when Western DDT bans had cost so many lives.

Like TASSC's other campaigns, the DDT campaign drew on the expertise and media connections of a cadre of professional science deniers who had been denying and distracting from science unfavorable to industry for years. Tobacco companies funded a neoliberal economist who had previously written dismissively of secondhand smoking's harms to publish a new "account of the DDT controversy." And they funded Bate, whose ESEF had previously denied the harms of secondhand smoke, endocrine disruption, and global warming. Now Bate authored a document titled "International Public Health Strategy," which wove its way through tobacco executives' inboxes as the POPs convention brought DDT back into the public eye.

Bate's strategy pulled threads from the scientific debate about DDT and spun them into a tale that warned against Western-led public health. His strategy echoed Attaran's letter and then added a nefarious spin. Malaria rates were climbing globally, he wrote, especially in Africa, and decades of epidemiological research on DDT had failed to turn up conclusive evidence of harm to health. The link to cancer, in particular, was weak at best. It was time, he said, to amplify the idea that environmentalists' unfounded vilification of DDT had placed millions of young, poor children at risk of deadly infection. DDT wasn't just another example of "junk science," according to Bate. A revision of its history would accomplish what few other stories about science, health, and the environment could. "You can't prove DDT is safe, but after 40 years you can't prove it's guilty of anything either," he wrote. Yet DDT had remained "such a totemic baddie for the Greens" that if you could pin a moral dilemma to it, it would pit liberals loyal to the environment against those devoted to public health. It was, he argued, an issue "on which we can divide our opponents and win."

The tobacco companies appeared to have been convinced. Bate collected 50,000 to 150,000 British pounds in payments from British American Tobacco and fees of 10,000 pounds per month from Philip Morris's Europe offices. He and his ESEF staff set to work publishing op-eds, books, and fact sheets on DDT's benefits and the ban's harms. And the argument gained momentum.

"It's time to spray DDT," wrote popular columnist and author Nicholas Kristof. "DDT killed bald eagles because of its persistence in the environment," wrote editorialist Tina Rosenberg in the *New York Times*, and "*Silent Spring* is now killing African children because of its persistence in the public mind." *ABC News* reporter John Stossel wondered how else environmentalists had misled the country. "If they and others could be so wrong about DDT, why should we trust them now?" he said.

The tobacco companies were pleased. "Bate is a very valuable resource," said one Philip Morris executive. "Bate returned value for money," said another.

Bate didn't act alone. The Competitive Enterprise Institute (CEI), a think tank whose scholars had spent the nineties defending tobacco and denying global warming, launched a website, www .RachelWasWrong.org, featuring the school photos of African children who had died of malaria. CEI's site said its partners included the ACSH and an equally anodyne-sounding organization called Africa Fighting Malaria.

On the Malaria Foundation International website, where Attaran and Galinski had posted updates on DDT's fate in the POPs treaty talks, a note appeared. Now that the DDT exception had been secured and the POPs treaty had been signed by most of the parties present in Johannesburg, the organization was moving on from its DDT campaign. "Further important progress is now being carried out by the Africa Fighting Malaria organization," MFI's website announced. A link took visitors to Africa Fighting Malaria, which described itself as a "non-profit health advocacy group." Its board chair was Bate. Its core staff of three included a woman named Lorraine Mooney, a close associate of Bate's who had previously run the ESEF. And its funders included foundations and think tanks promoting free-market ideals and Exxon Mobil.

With scientists moving on from the DDT issue, free-market defenders picked it up and ran with it. And people who bought DDT's new story as they came across it on the fast-growing internet took it from there. Before long, websites, blogs, and chat rooms were filled with people calling Rachel Carson a "paranoid liar," "mass murderer," and worse. She was responsible for more deaths than Adolf Hitler, they said. Dead more than forty years, she once again became for conservatives a potent symbol of the hazards of liberalism.

•

Larson had her second child at the end of 2002. By then, she had an international reputation as an advocate for tribal sovereignty and a cleaner Arctic environment. She traveled frequently to attend conferences and give talks. ACAT had sent her to Geneva and Washington multiple times. She had meetings with Alaska senators Ted Stevens and Frank Murkowski. She was the youngest member of her tribal government, the Chickaloon Traditional Village Council. A progressive magazine named her a "young visionary," one of "30 under 30" who were changing the world. "Becoming involved that young on international treaties is not that common," the magazine's editor said.

Larson had grown only more committed to her cause since her trip to Johannesburg two years before, especially because the US had signed but not yet ratified the POPs treaty. Increasingly, her world was full of disconnections and bad choices. She believed in breast-feeding her newborn, even as she now knew that this practice transferred chemicals from her own body into her children's. She wanted her children to grow up eating traditional foods, even as she knew they contained the same chemicals. She spent enough time talking to scientists to know that they didn't always understand her community's reality. Alaska Natives might eat fish twelve or fifteen meals a week, not two or three. When people saw runny bone marrow in moose and caribou, lesions on fish, and seals and bears skinnier than they used to be, they wondered if the animals' illness was connected to their own. Cancer was rare among Alaska Natives fifty years earlier, but in the nineties it had become the top cause of death. "We see things our elders never used to see," Larson said. She wondered if Alaska Natives were the "canary in the coal mine" for the rest of the world.

Elsewhere in Alaska, a woman named Paula Easley published a commentary in the *Anchorage Daily News*. "We now have West Nile Virus in America," she wrote. "Congress should revisit DDT's suitability for public safety uses." The environmentalists behind its ban, she said, should never have been trusted. She quoted Charles Wurster saying that people dying of malaria and other diseases was a good way of getting rid of them—it was the rumored and disputed quotation printed in the *Congressional Record* decades before, and Bate's ESEF had just published it in a recent book. She also quoted Bate saying that banning DDT was immoral.

West Nile virus hadn't appeared in Alaska, but Easley raised its specter anyway. "I shudder to think of the health consequences to Alaskans of a mosquito-borne virus epidemic, especially considering our endless wetland breeding grounds and the annual migration of potentially infected birds," she wrote. The bio at the end of her op-ed said she was a public policy consultant. It didn't mention that she was also on a select list of TASSC supporters.

•

Three years later, Larson flew to Uruguay for the First Meeting of the Conference of the Parties of the Stockholm Convention, the convention's governing body. The treaty was in full effect, ratified by dozens of countries by then—but still not the United States. To ratify the treaty, Congress first needed to amend and modify the existing federal laws governing the treaty's targeted pollutants, which included FIFRA and the Toxic Substances Control Act, a 1976 law regulating industrial chemicals. Several members of Congress had introduced relevant bills, but none of the bills ever made it to a vote.

So the US was relegated to observer status at the meeting, called to establish procedures for settling disputes between parties. It made little difference to Larson, who spoke on behalf of the International Indian Treaty Council. She wanted to make sure that traditional knowledge from Indigenous communities guided the convention's decisions and actions. She also wanted delegates to know about the forces working against them. "The chemical industry is trying to undermine the inclusion of new chemicals" in the treaty, she said. It was one more reason that the US Congress was unlikely to ratify the convention anytime soon.

The participating delegates worked through a long list of agenda items, including the exception for DDT. Ratifying countries agreed to notify the secretariat if they intended to produce or use DDT. If they did, they agreed to explain the disease-management strategy it was deployed in. They agreed to disclose how and where it was stored, and to let the secretariat know whether it was kept under security, under a watertight roof or not, and whether it was leaking into the environment in any way. When they agreed on procedures for settling disputes as well, they added them as an "annex" to the text of the treaty, to take effect in 2007.

Bate and his organizations and partners kept up a steady stream of commentary on the DDT issue all the while. He told journalists that the treaty's reporting requirements were onerous. He said the push to ban DDT in the first place was an "obvious example of eco-imperialism." In the long wake of the terrorist attacks on the World Trade Center in New York, he took to equating the number of African deaths from malaria to "a September 11 every 36 hours."

The year after the Uruguay conference, 2006, both the US Agency for International Development and the WHO issued formal statements on DDT for malaria control. In truth, the statements were readjustments of their respective long-standing positions on the pesticide. But they were important political statements at a moment in which Western involvement in the health affairs of less wealthy nations had been successfully recast as imperialist and hypocritical. USAID's statement clarified that the agency hadn't ever banned DDT; it just hadn't prioritized it. The WHO said that it had never stopped recommending DDT but that it had urged countries to use malaria drugs and insecticide-treated bed nets instead. Now, however, it was adding DDT back to its frontline defense against the disease. "We must take a position based on the science and the data," said the organization's malaria program director. "One of the best tools we have against malaria is indoor residual house spraying. Of the dozen insecticides WHO has approved as safe for house spraying, the most effective is DDT."

The statement was a clear message to countries that had banned DDT but were still grappling with malaria, like Tanzania and Uganda. They lifted their own bans or restrictions on the chemical. By the following year, global DDT use, which had declined from the seventies through the nineties, had doubled from year 2002 levels. In the meantime, West Nile spread to forty-two US states and Washington, DC. Like New York in 1999, they sprayed. This time, many cities and counties used pyrethrins, insect-killing compounds derived from chrysanthemum flowers, or pyrethroids, synthetic versions of the same, with power-boosting chemicals mixed in. "We use the safest material possible," a California official said.

Reporters soon caught on to Bate's game, tracking down documents he had sent to the tobacco companies and poring through his organizations' tax records. He was a "clever, slippery and often triumphant adversary," concluded one reporter, "a free-market Wizard of Oz, pulling levers behind the scenes."

Bate came to his own defense in an article in *Prospect* magazine. He said he had proposed the idea of a pro-DDT campaign because he thought it was "scientifically valid" but that the tobacco industry never funded it. (Documents contradict the claim.) Tobacco companies opposed DDT, he said, because they didn't want it contaminating their product. (The argument was true but out of date.) If scientific opinion on DDT shifted, he said, it was because of a letter signed by hundreds of malaria scientists and three Nobel laureates. He had been in Johannesburg, he said; he saw people change their minds in response. He didn't give Attaran credit for the letter.

Attaran, by then, was a professor at the University of Ottawa, where he held a prestigious research chair and had a new cause: railing against the "foreign aid industrial complex." A reporter from the journal *Nature* accompanied him on a drive through Ottawa to profile him. His colleagues, she said, called him a "likeable rascal." By the "tender age" of thirty-nine, he had run a successful campaign to protect DDT from a global ban, accused the WHO for failing to back more-effective malaria drugs, and lambasted the USAID for wasting money and obscuring its malaria spending.

"It's heartbreaking when you see people in the foreign aid establishment travel on a business class ticket to another country, check into a four-star hotel, and go to a conference on how to help the poor," Attaran told her. "I've done it enough, so I'm culpable." Presumably, he had changed his ways.

Chapter 28

TIMING MAKES THE POISON

May 27, 2007, would have been Rachel Carson's one-hundredth birthday, had she lived that long.

Angela Logomasini, a Fellow at the CEI, marked the occasion with an op-ed. Congressional Democrats were introducing legislation to commemorate Carson, she wrote, but they didn't know the truth about her legacy. Carson's work was "weak," she said, and full of "junk science." Because of Carson, hundreds of millions of people, mostly African children, were dying. Carson's true legacy, said Logomasini, was "misinformation." "Why should we—and our tax dollars—honor that?" she asked.

In the "Bring Back DDT" era, Carson was cast by free-market ideologues as the primary figure behind a deadly ban. And one of the top pieces of evidence that she was "wrong," in their view, was her claim that DDT and other pesticides caused cancer. "Cancer is caused primarily by smoking, poor diets and infections," Logomasini wrote, echoing a decades-old ACSH talking point. "There is little evidence man-made chemicals are causing a cancer epidemic." Carson's own breast cancer, meanwhile, became—as Carson had long ago feared it might—proof that her thinking was clouded and her conclusions foregone.

Logomasini wrote her attack on Carson not long after the CDC updated its "toxicological profile" of DDT and its main breakdown products, DDE and DDD, a report mandated by a 1986 amendment to the law that had created Superfund. The report, which summarized decades of scientific studies of DDT's effects on animals and people, stated that in people DDT seemed linked to several reproductive effects, including shortened lactation time, preterm births, decreased fertility, and low birth weight. Countless studies had also examined its relationship to cancer, especially breast cancer, but the results there were far more mixed. Numerous studies had found an association between DDT or DDE levels and cancer, but nearly just as many found no relationship between the chemical and the disease. And a good number of cancer studies were designed in ways that made definitive conclusions just about impossible, either because they didn't include control groups, for example, or because the people in them were exposed to multiple chemicals, not just DDT. "Overall, in spite of some positive associations for some cancers within certain subgroups of people," the report concluded, "there is no clear evidence that exposure to DDT/ DDE causes cancer in humans."

This scientific uncertainty, inherent to so many epidemiological findings, formed the basis for the conservative attack on Carson. But so did the scientific focus on cancer. Scientists were studying DDT's other health effects—on attention, memory, body weight, and endocrine function generally, for example—but postwar cancer fears and then the breast cancer movement had made cancer, and especially breast cancer, a top focus of scientific investigation. And that focus, by the early twenty-first century, gave DDT's defenders a place to stake their claim. It also led to easy dismissals of the chemical's toxicity in presumably more-objective circles. The claim that DDT causes breast cancer, reported the journal *Nature*, was "debunked."

Yet even in places where DDT had been used only minimally, as an indoor spray to control malaria, it was still turning up in people's blood and women's breast milk. And by the early 2000s, scientists had determined some of the more precise hormonal effects of DDT's isomers and breakdown products. The o,p' isomer of DDT was strongly estrogenic, for example, while p,p'-DDT was antiandrogenic, meaning that it blocked the activity of male hormones, whether synthetic or natural. The fact that these hormonally active compounds were still in women's blood and breast milk inspired a few scientists to look at their effects on the neurological development of newborns.

One was a University of California, Berkeley, professor named Brenda Eskenazi, who for several years had been conducting an ongoing study of the effects of pesticides on a cohort of children born to mothers in California's agricultural Salinas Valley. Many women in the area were recent immigrants from Mexico, which had only recently banned DDT, in 2000. Eskenazi and her colleagues had blood samples from women who were pregnant, and they had these analyzed for DDT and DDE levels as they followed up with their newborns, in order to track their development. They used a standardized test to measure the infants' cognitive, language, and motor development, testing to see whether a six-month-old could bang a toy, for example, or whether an older infant could babble or engage with an adult. Overall, they found that the higher the mothers' DDT levels, the lower the babies' test scores. In utero exposure to the chemical, they concluded, might be slowing or harming the neurodevelopment of young children.

But they also found that breast-feeding seemed to counter the effects. The more a woman breast-fed, the better her infant's scores, even if she carried a higher level of DDT in her body.

Eskenazi was circumspect about the findings. Even with a potentially negative effect on infant development, she said, in places where malaria was widespread, DDT's risks still paled in comparison to its benefits. She put it bluntly. "I'd rather have a child with three IQ points less than have a dead child," she said.

•

At the National Cancer Institute in Washington, Jennifer Rusiecki, one of the institute's postdoctoral researchers, was puzzling over what to do with a data set she had been handed. She got the data from another postdoc, who had just moved on to a permanent job, like all of the postdocs usually did after a year or two. That postdoc had gotten the data from Susan Sieber, the National Cancer Institute scientist who, back in the nineties, had thought it would be a good idea to study the rate of breast cancer among women in the Triana Medical Fund.

Sieber had spent the late nineties getting the study off the ground. She teamed up with a few faculty members from the nursing school at the University of Alabama in Huntsville, and at the Triana Medical Fund's annual health fair they had invited women to participate. It took a few years, but eventually they had blood samples and mammograms from 230 women in the area. Then Sieber herself was diagnosed with breast cancer. In 2002 the disease took her life. The Triana data, of course, remained. And now the data belonged to Rusiecki.

As Rusiecki looked the data set over, she realized she'd need to think creatively about what to do with it. There weren't enough women in the study to draw any conclusions about breast cancer, she realized. The disease was common, but not common enough that she'd be able to extract meaningful findings from a sample of people that small. Gradually, she realized that she could study the women's breast health instead. She designed an analysis that used the women's mammogram data to see how common markers of breast disease, such as breast-tissue density, were among them.

Then, like all of the other NCI postdocs, Rusiecki got a permanent job offer somewhere else. She accepted the offer, made plans to move, and decided to take the data with her. She knew she'd be busy at her new job, teaching students and carrying out new research. But someday, she figured, she'd find a way to get back to it.

•

Just a few blocks north of Eskenazi's lab, in a little office just past the edges of the Berkeley campus, another epidemiologist, Barbara Cohn, realized that she also had data that might shed light on DDT's health effects—especially its much-disputed connection to breast cancer.

Cohn, a quick-thinking, always-moving mother of three, worked for a nonprofit research institute where she had also inherited a data set in a roundabout way. The "data," originally, was some twenty thousand placentas and vials of blood serum taken from women who had given birth at the Kaiser Permanente Hospital in Oakland, California, between 1959 and 1967, collected by a longtime Berkeley statistics professor. When the professor died, he left the data to one of his research collaborators, a Dutch epidemiologist named Bea J. van den Berg, who worked in Berkeley's School of Public Health. Van den Berg expanded the data set by inviting the women's children to join the cohort, and then she collected samples from them. When van den Berg lost her position on campus, she found a place for herself in Cohn's office—and a new collaborator in Cohn, who saw the limitless research potential in the samples. When van den Berg retired, she left them in Cohn's care.

As an epidemiologist, Cohn was a bit of a generalist; she had studied twins, children's health, heart health, and cholesterol, as well as cancer. Watching the national debate over DDT unfold, she realized that the samples in her charge might hold answers to its health effects. She wasn't terribly invested in the outcome either way. She didn't see herself as "some flaming environmentalist," she said. "I was just curious."

Cohn knew that because the blood had been drawn from women in the 1950s and 1960s, it was like a time capsule. The women, for the most part, were still alive, but the blood samples showed what their DDT levels were when they were young: just girls or teenagers. That is, the samples were an artifact from a time when DDT was still sprayed widely on farms and forests and was still ubiquitous in the food supply. To learn from them, she needed to find someone who knew how to detect the pesticide without using up too much of the serum. A colleague put her in touch with Mary Wolff at Mount Sinai, who agreed to help. With two other researchers, they designed a nested case-control study. They identified 129 women in the cohort who had developed breast cancer by the age of fifty, and then they compared their DDT levels to cancer-free women in the cohort who were born in the same year.

Their finding was astonishing: women who were exposed to DDT before the age of fourteen had a fivefold increase in breast cancer risk. The younger they were when exposed, the higher their risk.

It was easy for Cohn and Wolff's study team to explain why no other study had found as strong a connection between DDT and the disease. Theirs was the first to measure DDT, not its metabolites, in blood samples taken when the cases were young and the chemical's use was still high. The chemical, they hypothesized, might have damaged DNA or induced enzymes that produced compounds damaging to genetic material. It was possible, they ventured, that the breast was vulnerable to DDT's cancer-promoting effects only when women were young and their bodies were still developing. Their finding, they wrote, was "consistent with results obtained in studies of exposure to atomic bomb radiation." There, too, breast cancer risk rose in women who were young when exposed.

The study was also a powerful validation of Colborn's hypothesis, in people. It wasn't necessarily just the dose that made the poison. Sometimes it was the timing.

•

By 2008, 162 countries had ratified the Stockholm POPs Convention, a number that made it one of the most successful international agreements ever to protect the environment. DDT, in keeping with the treaty's exception, was still in use in several countries in Asia and Africa. It was manufactured in just three countries: India, China, and Russia. And wherever it was used—even in small amounts as an indoor spray—it continued to appear at high levels in people's blood.

That spring, Alma College, just north of the Pine River in central Michigan, hosted an international conference of DDT experts. The college was downriver from the onetime fifty-four-acre home of the Velsicol Chemical Corporation, which had purchased DDT manufacturer Michigan Chemical Corporation in the 1960s. In succession, the companies had produced DDT, flame retardants, and other chemicals on the site from the 1930s through the 1970s. In the 1980s the acreage had become a massive Superfund site. Since the 1990s, the EPA had been overseeing its cleanup. But in the 2000s, DDT and other chemicals were still leaching from the site and into the river.

Local residents had long since organized themselves into a task force to stay on top of the cleanup's progress. At one of their early meetings, a task-force member had brought in a copy of a *Wall Street Journal* op-ed that had caught his eye. "A DDT Ban Would Be Deadly," its author, a demographer with the ESEF in Cambridge, England, argued. Task-force members were stunned. "We've been fighting to clean up DDT and now they're saying it's fine?" said member Ed Lorenz, an Alma professor of history and public affairs.

With Lorenz as a liaison, the task force turned to faculty and students at Alma for help organizing a conference that would bring together scientists who were studying the effects of DDT on people and the environment. Brenda Eskenazi accepted their invitation to speak. So did Barbara Cohn, along with about a dozen other scientists who had published watershed studies on DDT. The task force welcomed them to town, gave them a tour of the Superfund site, and listened intently as they each presented their own scientific work.

At the end of the conference, Eskenazi made a suggestion. "We can't all go home," she said. "We have to do something." The assembled scientists agreed that global policy discussions would benefit from their own review of the latest research on DDT, especially because so many studies had been conducted since the POPs treaty was first signed. Together, they agreed to review nearly five hundred studies of the chemical. On balance, the studies suggested that DDT might be associated with breast cancer, impaired neurodevelopment in children, diabetes, spontaneous abortion, and decreased semen quality in men. DDT had value, they conceded, but it still needed to be approached as a solution of last resort.

"We cannot allow people to die from malaria," said one of the scientists, a public health professor from the University of Pretoria in South Africa. "But we also cannot continue using DDT if we know about the health risks." They titled their summarized findings—published in the same journal that had published Colborn's endocrine-disruption paper—the "Pine River Statement: Human Consequences of DDT Use."

By the time the statement was published, the EPA's cleanup efforts had removed hundreds of thousands of cubic tons of sediment and soil containing DDT. For communities living near the old Velsicol site, however, DDT wasn't their biggest contamination problem. The site had shut down in the seventies because of its role in one of the worst chemical poisonings in US history. A plant had accidentally mixed a fire retardant into animal feed, and the contaminated feed was sent to farmers across the state. Nine million people had consumed the poison, which mimicked estrogen and accumulated in body fat, causing thyroid problems, miscarriages in women, and slowed growth and early puberty in boys and girls.

Now, the Velsicol site was one of the biggest and most expensive Superfund sites in the country. In one part of the cleanup, the EPA's contractors dug up DDT-contaminated soil from the yards of a neighborhood bordering the site. They dug so deep they created temporary moats around each home. When they believed they had dug far enough, they installed plastic liners and placed new clean soil on top. Five years later, however, homeowners said they were finding dead and trembling birds in their yards.

The task force asked a wildlife toxicologist from Michigan State University, Matthew Zwiernik, to test them. The DDT he found concentrated in the birds' brains was ten times the level known to be fatal. Zwiernik said the level was higher than anything he could find in published scientific papers. "I was shocked," he said. "It's like nothing I've seen in my career." It was 2013, and in central Michigan, it appeared, DDT was still killing birds.

The local task force, which paid for Zwiernik's study, meets monthly to this day. The EPA still warns against eating fish caught near the site. Its contractors are studying new ways to remove DDT from soil, where scientists now know it can last for dozens and potentially hundreds of years.

•

Both Eskenazi and Cohn came back to the DDT question again and again after the conference in Pine River, each in their own way.

Cohn kept digging into the samples she inherited to see what else they might reveal. Eskenazi launched a new cohort study, recruiting more than seven hundred pregnant women in one of the hottest districts in Limpopo province, in the northernmost corner of South Africa, to follow over time. Malaria is common in the Vhembe district, especially in the rainy season, when mosquitoes can breed in the water puddled in a single hoofprint. In the province as a whole, hundreds of people contract the disease each day.

Sub-Saharan Africa was—and is—one of the last places on Earth where DDT is still used. Made by just one company in India by the 2010s, it had become precious to malaria-control officials. They came to see it as gold, just as Robert Mizell had eighty years before in Georgia. They sprayed it inside people's homes, often in combination with other pesticides, because it was one of the cheapest and most effective tools at their disposal.

For the past decade, Eskenazi and a former student of hers, who is now a professor at McGill University, have been measuring the amounts of DDT and other pesticides in the homes and bodies of women and children in Vhembe. They've found that women with higher DDT levels had a higher risk of pregnancy-related hypertension. Their infant daughters were heavier and bigger for their age, their sons had lower levels of a thyroid hormone responsible for growth and development, and in general their children had more symptoms of allergies, including wheezing and rashes, and more frequent fevers than the children of women with lower levels of DDT.

And that was with careful use of the pesticide. Spray teams were trained to make sure that DDT didn't drip from thatch-roof eaves onto the ground, or onto children's toys, or into open water receptacles. They told people to put away their kitchenware and utensils before the spraying. They sprayed just a small, "residual" amount, and only indoors. The pesticide still got into people's bodies. And it still affected their health. Yet Eskenazi came to think that didn't mean there wasn't a place for DDT where malaria threatens. "I'm not a radical," she said. "There are babies dying. If the baby's dead, there aren't going to be long-term effects."

Around the same time that Eskenazi launched her study in Limpopo province, Cohn conducted another nested case-control study, this time looking at breast cancer in the daughters of the original women in the study. The daughters, they found, had a fourfold increase in breast cancer risk if their mothers had higher levels of DDT while pregnant. A few years later, she and her colleagues looked at two risk factors for breast cancer—obesity and early menstruation—in the women's granddaughters. Once again, the higher the grandmothers' DDT levels, the greater the granddaughters' risk. In about a dozen studies, Cohn and colleagues found these and related effects again and again. At conferences of epidemiologists and environmental health scientists, people would approach her "with their mouths hanging open," she said. "No one gets this scale of effect."

The head of the National Institute of Environmental Health Sciences, part of the NIH, called Cohn's research cohort "a national treasure that keeps on giving."

•

In the meantime IARC, the WHO's cancer agency in France, convened a working group of two dozen of the world's top DDT experts to review the latest research on the chemical and reassess its carcinogenicity, as it had done every ten years since 1975. On a regular basis, the agency reviews the carcinogenicity evidence on hundreds of substances and materials. At the DDT meeting held in 2015, the agency confirmed the pesticide as a "probable carcinogen," one step away from a definite carcinogen. The classification meant that evidence of DDT's ability to cause cancer was "strong"—just not absolutely conclusive.

The agency's ensuing report, or "monograph," emphasized DDT's connection to testicular cancer, liver cancer, and non-Hodgkin's lymphoma, as demonstrated by studies conducted in the US, China, Norway, and elsewhere. It stated that there was no connection between DDT and breast cancer. It included a nod to Cohn's work, acknowledging that "the possible importance of early-life exposure to DDT remains unresolved."

The report made barely a ripple, however, in part because that same year IARC was embroiled in controversy over some of its other working groups' findings. One had declared red and processed meats carcinogenic, pointing out the latter's strong link to colorectal cancer. Another had found the pesticide glyphosate to be probably carcinogenic, even as an EPA review had concluded that it was safe. Critics accused it of conflict of interest and lack of transparency. Some of its most outspoken critics, unsurprisingly, included (and still include) groups such as the American Chemistry Council.

But its critics also included scientists who had taken part in the agency's reviews, under confidentiality agreements, and said that factors such as human bias, ego, and reputation often colored its conclusions, because they could determine which studies the working groups considered and which ones they did not. With so many studies on DDT and breast cancer, all of those factors certainly influenced the findings the working group took into account. But those confidentiality agreements prevent members from saying as much openly.

•

Scientists all over the world have studied—and continue to study—the relationship between DDT and a list of health effects: obesity, diabetes, multiple forms of cancer, Alzheimer's disease, Parkinson's disease, and more. Few researchers have time capsules like Cohn does. More have designed so-called prospective cohorts, like Eskenazi and her colleague have. Far more are so-called cross-sectional studies, which look at the relationship between an exposure (such as DDT) and an outcome (such as diabetes) at a specific moment in time. Because they look at just one moment, they reveal in-the-moment relationships, but not relationships that prove causality.

Yet taken all together, the body of DDT studies available today sheds new light on the pronouncements of more than a decade ago, that "Rachel was wrong" and DDT's harms were overblown. A few years ago, a philosopher and a biologist at Washington State University argued that DDT's use is now a matter of transgenerational environmental justice because it causes harm to future generations that had no say in past decisions to use it. "Those who might live in the future are the ultimate socially disempowered group," they wrote.

Akilah Shahid, a thirty-one-year-old Early Start case manager who lives in Walnut Creek, California, knows exactly what that feels like. Her grandmother was a nurse who had spent her career working at Oakland's Kaiser Hospital and who had battled cancer several times. Several years ago, Shahid learned that her grandmother was part of Cohn's study—and that she and her mom were in the study, too. Suddenly, so many things made sense to her.

Shahid was a biochemistry major who taught high school biology after college, and she had always carried a vague memory of learning about DDT in her own science classes. Now, it seemed to be the reason why she had been one of the first girls in her middle school to get her period; it also seemed to explain why she'd always struggled with her weight, and why her grandmother's sister also had cancer—breast cancer. "It was shocking," she said, to learn that something trusted in her grandmother's time had come to shape her life so profoundly. "You think you can actually trust the people who know, but you realize you can't."

Now she participates in an advocacy group that tries to convince people to reduce their exposure to endocrine disruptors by avoiding pesticides in food, plastics, and chemicals found in everyday items such as receipts. The group's larger aim is to reduce women's risk of breast cancer. Shahid said she is more aware than ever that everyday chemical encounters may be sealing the fate of a future generation, just as they long ago sealed hers.

"Why do we keep being exposed to things?" she said. "We say that's tomorrow's problem, but tomorrow's now."

•

Angela Logomasini is still a Fellow at the CEI, where her specialty is environmental risk: she writes and talks on television and radio about the "alarmist" nature of scientific studies linking chemicals and health problems. She continues to dismiss "junk science."

Her "Safe Chemical Policy" website, hosted by the CEI, highlights "Rachel Carson's dangerous legacy" and dismisses the risks of DDT as a "cultural myth." The website's partners include the conservative Heartland Institute, Americans for Tax Freedom, and the ACSH.

In the weeks leading up to the Capitol insurrection on January 6, 2021, Logomasini took to Twitter to spread messages—some might say misinformation, or even disinformation—about alleged electoral fraud behind Donald Trump's loss in the 2020 presidential election. "Don't trust Congress! Don't trust bureaucracies!" she tweeted. "They will shut down free speech permanently. Don't fall for it."

Progressive bloggers captured her tweets and published a profile on her, detailing her denial of chemical and environmental risks. By the end of the month, her Twitter account was gone.

Her "Safe Chemical Policy" site is still up.

EPILOGUE

Clyde Foster's daughter Edith spent most of 2020 in lockdown in Triana, taking care of her adult son and her mother, who lives next door. We "basically hunkered down," she said. She watched as members of her extended family came down with COVID and struggled to recover, some with better luck than others. As a doctor, she said, she "never imagined that the immune system could be turned on so badly by a virus."

She got herself, her mom, and her son vaccinated early in 2021; one of her siblings enrolled in a vaccine trial. Later that spring, with COVID in temporary retreat, she went down to Indian Creek for a party to celebrate a grandnephew's graduation. She hadn't been there in ages, and not because of COVID. "I'm not an outdoors person," she said.

The river was beautiful, glassy and still, its banks lush and the surrounding trees full of life. Boat ramps led in to its depths. Signs invited visitors to fish. From a nearby parking lot, a glossy wooden sign pointed the way to the Foster Loop Trail, named for her father, who died in 2012.

That same year, the Alabama Department of Health lifted its restriction on consuming fish from the area. For the first time, their DDT levels were low enough to be considered safe—in areas of the river open to the public. The Superfund site, still under EPA's watch, is divided into three areas, or reaches, and the most contaminated reach, Reach A, is the stretch of tributaries closest to the old DDT plant on the Redstone Arsenal. Under the terms of the cleanup agreement, DDT levels in fish must measure less than 5 ppm for three years in a row for the cleanup to be considered complete, and the last of the monitored fish species, smallmouth buffalo, met that goal only in 2015.

But that wasn't the end of the cleanup in Triana, by far. The consent decree requires that all fish be sampled seven years after the last one meets the target DDT levels, to ensure that levels have stayed low. That seven-year test will happen in 2022. But in its sixth five-year report on the cleanup, released in July 2020, the first summer of COVID, the EPA noted that the target DDT levels might not be low enough to protect wildlife or human health. Fish-eating birds could still face risks from the current DDT levels in fish. And people eating thirty-five half-pound fish meals a year would face an elevated cancer risk. The DDT levels would need to be reduced to 2 ppm, not 5 ppm, to diminish that risk, the report said.

Scientific knowledge, not surprisingly, has changed since the decree was signed in 1983.

•

Jennifer Rusiecki, the former NCI postdoc who was handed the data from the Triana medical fair fifteen years earlier, also spent 2020 sheltering in place. As it happened, the month that the pandemic shut down so much of public life in the US was also the month that her analysis of the data was finally published. In the end, she and her collaborators had looked at the relationship between breast density and levels of DDT, PCBs, and six other chemicals. Breast density, an indicator of heightened breast cancer risk, was highest in women with high levels of five of the chemicals—but not DDT. Focusing in on the DDT metabolite DDE, they found that if women were young at the time of exposure, even with DDE levels correspondingly low, they nonetheless had higher breast density. Age of exposure once again seemed to matter to health—but it was difficult to draw firm scientific conclusions about the relationship between the two. She and her collaborators had data from slightly more than two hundred women between ages nineteen and ninety-one, which meant that the number of women in any specific age range was small.

In an echo of the first CDC study of Triana residents more than forty years before, Rusiecki and her colleagues found that the women in their study had among the highest levels of DDE in the world. Yet the relationship between PCBs and women's breast density was much stronger—especially in women who were young when chemical contamination in Triana was at its highest. The source of the PCBs in the river remains a mystery, which means some residents worry that it may return. Before her and her colleagues' study was published, Rusiecki, now a professor at the Uniformed Services University of the Health Sciences in Bethesda, reached out to Marvelene Freeman. The pair made plans for Rusiecki and Tom Mason, now at the University of South Florida, to come to Triana to share the findings in a community meeting. But because of COVID, the meeting was put off.

Up in New York, Mary Wolff, at Mount Sinai, saw the published study and wrote a letter to the editor of the journal it appeared in. Wolff was struck by the fact that Rusiecki and her collaborators had adjusted their findings for body mass index (BMI), meaning, essentially, that they removed data from individuals whose body weight was very far from average. To Wolff, that mattered because studies over the last decade have shown that obesity extends the half-life of persistent organic pollutants in the human body. It also mattered because studies have shown that age influences the relationship between body weight and half-life. The relationship between age and effect of DDE in Rusiecki's study, to her, pointed to something that stretched all the way back to Colborn's report. The time of exposure matters. "It's a critical window," said Wolff. She also acknowledged how difficult it is to draw connections between chemical exposures and health outcomes like cancer. The more scientists study the connection, the more complex it seems. "It's the window, it's the [chemical] mixtures, and it's how the environment interacts with different genes at different moments in time," she said. "Plus, we've moved on."

For despite the press for DDT's continued use in areas battling epidemic malaria, organochlorines, especially organochlorine pesticides, have been all but abandoned. Starting in the seventies in the United States, they were replaced by the organophosphate pesticides, which didn't persist but were acutely toxic to farmworkers and wildlife. The organophosphates' eventual dominance, said historian Frederick Rowe Davis, was "one of the most tragic ironies of the DDT ban." In the early 2000s, following safety evaluations required by the Food Quality Protection Act of 1996, the EPA began restricting the use of some of the most toxic ones. But the agency's pesticide actions are as politicized as ever: environmental groups turned to legal action to force the EPA to take one particular organophosphate, chlorpyrifos, off the market. The agency began taking steps to do so—and then, after the election of Donald Trump, reversed course. Less than a year after Joe Biden was elected, the agency reversed course again. As of early 2022, Biden's EPA announced, the pesticide will no longer be allowed on food crops.

And yet another new generation of pesticides has already taken over. Neonicotinoids were developed in the late 1980s and introduced in the 1990s, while lawmakers in Washington were hammering out new regulations to ensure greater pesticide safety. Today, neonicotinoids, or "neonics," are the most commonly used insecticides in the world. One of their appeals is that they are so highly toxic to insects that smaller amounts are more effective than much larger amounts of older pesticides. Another is that they're persistent: they remain effective for years. They're also used in targeted ways: applied as a seed coating, for example, or injected into tree trunks. The plants take the active ingredients up into their tissues. For this reason, the neonics are known as systemic insecticides.

Like other pesticides, neonics are neurotoxins; they interfere with nicotine receptors in insects' nervous systems. Over the last two decades, scientists have shown that neonics make their way into food webs, kill beneficial insects, and take a particularly hard toll on bees. Doses that don't kill bees outright can affect their ability to move, find food, and successfully reproduce. Bee populations are declining, and neonics bear a good part of the blame. The pesticides are toxic to other beneficial and nontarget insects, too. That, in turn, affects animals that feed on insects, including birds. "The full extent of environmental risk from neonicotinoids may not be known for years," warned a pair of scientists two years ago. "We risk a return to the toxic, simplified environment that prompted *Silent Spring*—a work that documents a history we should actively work to avoid repeating."

•

Six months after Rusiecki's study was published, readers in California, where I now live, woke up to a chilling story on the front page of the *Los Angeles Times*. Clicking on the story, the screen went dark. Scrolling down the page, a series of images zoomed in, gradually, on an old corroded barrel sitting on the ocean floor. The barrel was ringed with a phosphorescent glow that seemed to indicate where its toxic contents had leached out: DDT.

The story, a painstaking undertaking by a young *Times* reporter named Rosanna Xia, originated with a scientific paper published the year before. The paper's lead author, UC Santa Barbara Marine Science Institute professor David Valentine, had long ago come across a reference to Allan Chartrand's 1985 report describing barrels of Montrose's DDT waste dumped into the Southern California Bight. On a 2011 deep-sea expedition to study underwater methane deposits, he believed he had found one of the dump sites. He invited a new graduate student, Veronika Kivenson, to take on the project of sampling and analyzing sediment from the site. In 2013 they went to sea with a team of researchers and two submersible vehicles capable of capturing picture, video, and sediment from the ocean floor.

The other scientists on board their vessel weren't there to study the barrels; they had their own research missions. But on the last day of the expedition, they let Kivenson and Valentine take over the submersibles to scan the ocean floor. At Valentine's direction they guided the pair of them to the site he had identified two years before. Sitting before a wall of monitors with live feeds from the submersibles, Valentine and Kivenson saw dozens of barrels come into view. They stayed up through the night, directing the vehicles to collect samples. Months later, Kivenson and a colleague analyzed the material they had collected. Sure enough, the deep ocean floor contained deposits of DDT.

Kivenson became obsessed with the history of the dump sites. She went to county offices, tracked down old microfiche records, and took pictures of them with her cell phone. She went on eBay to find old books, long out of print, on how to ocean dump and where, and bought them for a dollar apiece. She dove into the scientific literature and found papers that researchers had published twenty and thirty years earlier on the DDT in the bight. She devised a way to determine if the DDT near the barrels, at a depth of up to 3,000 feet below the ocean's surface, was distinct from the DDT expelled from sewage outfalls closer to shore. It was. UCLA scientist M. Indira Venkatesan, independently developing and applying the same methodology, had drawn the same conclusion decades before.

When Chartrand heard about Valentine and Kivenson's findings, he felt vindicated. "I told you so!" he said.

It took Valentine and Kivenson six years to publish their findings. Xia's *Los Angeles Times* story turned the findings into front-page news. The passage of time had made the barrels a modern-day curiosity, shocking evidence of a once-common and now forgotten means of dealing with chemical waste. The barrels, as Chartrand well knew, weren't full of DDT, of course. From what he had learned decades before, they contained acid sludge from the manufacturing process, which usually contained no more than 1 or 2 percent DDT. Given the number of dumped barrels that Montrose had logged, however, that still amounted to hundreds of tons of DDT in the sea.

Other news outlets picked up Xia's story on the barrels. Xia raked in accolades and awards. California Senator Dianne Feinstein demanded that the EPA take action. Other lawmakers called hearings in the House of Representatives in Washington and the State Assembly in Sacramento. Foundations reached out with grant funds. Scientists convened meetings and conference panels. A team from the Scripps Institution of Oceanography and NOAA pulled together an expedition in record time to go down to the seafloor and map the dump site in the spring of 2021. They didn't see dozens of barrels this time; they saw more than 27,000.

By Chartrand's 1985 estimation, the total number of barrels may be closer to half a million. But the precise number may be beside the point. Kivenson and Valentine's samples showed higher concentrations of DDT in sediment collected farther from the barrels than those samples taken close to the barrels. It's possible the DDT waste was not just dumped in barrels but also poured directly into the ocean from massive holding tanks. It's also possible that much more was dumped than Montrose logged, in different places, and mixed in with chemical waste collected from other industrial sites. Either way, "There is a lot of DDT down there," Valentine said.

In a now-familiar pattern, an environmental group responded to the news of the deep-sea waste by sending Montrose Chemical Corporation and its current parent company, Bayer Corporation, a notice of intent to sue. The notice, given by the Oakland, California-based Center for Biological Diversity, cites DDT's human health effects, its harm to birds, and new evidence linking the chemical to a massive epidemic of cancer in California sea lions. Roughly one in five of the wild sea lions now has urogenital carcinoma. Studies of the marine mammals show that the risk of disease is highest in animals who were exposed to pollutants at a young age and then later infected with a form of herpesvirus. Pollutants and viral infection, researchers have concluded, appear to have what scientists call a synergistic effect in the species: they amplify each other.

But it's not clear what, if anything, can be done about the DDT in the bight. There is still DDT on land that has yet to be remediated. Cleanup efforts targeting the DDT disposed closer to shore, in relatively shallow water, were complicated, expensive, and not entirely effective. Dredging barrels from far-deeper waters would most certainly stir up deposits of the chemical, possibly making some forms of marine life even more likely to come in contact with it. Moreover, it's next to impossible to make cleanup plans when no one yet knows exactly how much waste is on the ocean floor, where it is, and what its toxic companions are.

At this point, only one thing is clear: seventy-five years after scientists first warned of its hazards, sixty years after Rachel Carson wrote *Silent Spring*, and fifty years after it was banned, DDT is still here.

•

DDT's story shows how long it can take to pin environmental and health effects to their chemical or technological causes. Decades after DDT was banned, it remains in people, fish, soil, and sediment deep in the sea. Scientific studies of its risks to human health and wildlife continue. As I wrote the history of this chemical, some of the science described in this book—on DDT's transgenerational effects in people, and its presence in and links to disease in marine mammals, for example—was just emerging. Some of the more recent science on the chemical seemed to uphold the patterns that emerged in the history that took shape as I wrote. I found myself unable to ignore the fact that United Farm Workers organizer Jessica Govea, exposed to so many pesticides as a little girl, died of breast cancer in 2005, at age 58, while Marjorie Spock, exposed relatively briefly and not until her forties, lived in good health to age 104. (Remarkably, Spock also farmed—on her own, privileged terms, it must be said—until she was 100.)

These are just anecdotes of course, stories that seem to mean something that scientists are still working to prove or dismiss. But they fit a broader pattern we are already well aware of: that historically the burden of environmental hazards and related health problems has fallen most heavily on those who are not wealthy or white. Today, countless epidemiological studies have shown as much. Of course, it's not a pattern we need epidemiology to be able to see. "Science is one way to know the world," said environmental activist Carolyn Finney recently. "It is a good way, a very good way, but it is not the only way to know the world."

Yet scientists are our "designated experts" for studying and understanding the world, notes science historian Naomi Oreskes. The sciences generally (some more than others, admittedly) began to accrue tremendous cultural authority during and after World War II. Spock acknowledged this authority as she gathered studies and reports for her trial. UFW staff grasped this as they asked the state health department to study farmworker health. So did Clyde Foster as he invited the CDC into his town. They, like countless others, believed that scientific knowledge, in the form of truths and facts, would yield justice in the face of breach of property, dignity, and health. It disappointed them all, in different ways. Its findings were open to interpretation, dismissed if not replicated, or judged too complicated to reproduce.

Of course, it was those very characteristics of scientific knowledge, combined with its cultural authority, that made an attack on science so useful to the Madison Avenue public relations executives working for the chemical and tobacco industries. They were deviously creative in devising ways to exploit the need for verification and refutation in science, as well as exploiting science's inherent uncertainty. By now, a substantial list of historians, journalists, and former government and industry insiders have documented the behavior. "Like McDonald's, Coca-Cola and Levi's jeans," wrote journalist Paul Thacker, "scientific disinformation is an iconic American product."

DDT's three-act story adds to the case studies that Thacker and others have compiled, and it complicates them. Industries manufactured doubt about science generally to protect markets for their own products. They sat back as regulation went their way and stepped forward when it didn't. And they shared common cause with the free-market defenders, such as Elizabeth Whelan and Roger Bate, whose communications wins built on decades of their own corporate doubt mongering. By now, the web of disinformation that such actors have woven is so thick it's impossible to see the original connections. "The coronavirus has largely granted environmental hypocrites their wish," wrote a Heartland Institute fellow early in the COVID pandemic. As evidence of environmentalists' radicalism, elitism, and hypocrisy, he quoted EDF founder Charles Wurster purportedly "praising the ban on DDT because it would result in millions of . . . deaths from malaria in developing countries." The 1940s through the 1970s gave us DDT in soil, water, and bodies; the 1970s through the 1990s gave us DDT as supposed proof of liberalism gone wrong. In the 2020s, both are still with us.

Today's news stories and social media threads often lament that science is in crisis, pointing to public doubts about climate change, vaccines, and the very nature of COVID. Decades of intentionally sowed doubt, along with other corporate and free-market practices, are certainly responsible for contemporary skepticism toward science in some circles. But that version of the story is too neat and simple; there's something else in the picture too. Science is a social process carried out by communities of people beholden, as all people are, to their respective relationships, communities, interests, beliefs, worldviews, and biases. It's infused with human values—including some of the most pernicious ones, such as paternalism, sexism, and racism. Doffie Colson certainly felt this as she sought help from scientific experts who dismissed her lived experience; Clyde Foster felt it too. Again, these are anecdotes, but they support another all-too-familiar pattern. "One cannot continue to breach trust," said bioethicist Yolonda Y. Wilson, "and then get upset when people don't trust you."

Oreskes and Wilson, among others, argue that there may be a few things we can do to preserve the value of science for all. We can, for instance, accept that science is social, that it grants some human worldviews (even harmful ones) an extraordinary degree of authority, and that it is just one way of understanding the world. We can also accept that the fact that so many scientific ideas have ended up, as Oreskes notes, "on the scrap heap of history" is not proof that science is bankrupt but is simply part of the scientific process—the very human and communal process we've agreed to use to understand and, as Oreskes puts it, engage with our world.

All of this risks turning DDT's story into yet a different sort of morality tale. More than that, however, it's an illustration: of forces unseen, of values unacknowledged, and of the endless game of catch-up we play when we pollute first, regulate later; deploy first, study later; and act first, reflect later. It's also about the very simple sum total of all of the stories herein, which Foster captured so plainly while talking to a reporter about the DDT in his town back in 1981. "The issue transcends Triana and Indian Creek and the Tennessee River," he said. "What we are talking about here is planet Earth. God only made one."

ACKNOWLEDGMENTS

The research and completion of this book were supported by the National Endowment for the Humanities, the National Institutes of Health's National Library of Medicine, the Science History Institute, Emory University, and the Graduate School of Journalism and the School of Public Health at the University of California, Berkeley. So many people gave me their time, thoughts, comments, expertise, assistance, documents, tips, advice, and support over the years that it took me to research and write, including the anonymous reviewers for *Environmental History, Southern Spaces, Public Health Reports*, and the National Library of Medicine, along with Charisma Acey, Jim Aidala, Geeta Anand, Leif Anderson (Department of Special Collections, Stanford University), Elsa Atson (Science History Institute), Amir Attaran, Ashley Augustiniak (SHI), Jennifer Barth (Wisconsin Historical Society), Peter Brown, Kristin Carden, Julia S. Chambers (Special Collections Research Center, Syracuse University Libraries), Kristen Chinery (Walter P. Reuther Library), Paul Civitelli (Beinecke Rare Book and Manuscript Library, Yale University), Barbara Cohn, James Colgrove, Carrie Crawford, Angela Creager, Frederick R. Davis, Elysa Dombro, Cathy Dorin-Black (North Carolina State University Special Collections Research Center),

Nancy Dupree (Alabama Department of Archives and History), Sandra Eder, Deirdre English, Reynald Erard (World Health Organization Archives), Brenda Eskenazi, Cheryl Fisher (California State Library), Moira Fitzgerald (Beinecke), Edith Foster, Eyal Frank, Marvelene Freeman, Margarete Froelicher, Emily Fuller (Sonoma County Central Information Bureau), Deena Gorlund (American Psychiatric Association Foundation), Terrence T. J. Greer, Guy Hall (National Archives at Atlanta), Marlene Harmon (University of California, Berkeley Law Library), Sean Hecht, Maureen Hill (National Archives at Atlanta), Barbara Hoddy (Arizona State University Archives and Special Collections), Abigail Li Holst, William Hays Hopkins, Bob Kenworthy, Brian Keough (Special Collections & Archives, University at Albany), David Kinkela, Uriel Kitron, Veronika Kivenson, Kathleen Kreiss, Howard Kushner, Michele La Merrill, Stephanie Lampkin (Jane and Littleton Mitchell Center for African American Heritage at Delaware Historical Society), Marsha Lash (Ashland County Historical Society, Ashland, Ohio), Ed Lorenz, Sonya Lunder, Tom Mason, Michelle Mart, Roland McDonald (Alabama Department of Archives and History), Aimee Medeiros, Abigail Meert, Michal Meyer, John Monahan (Beinecke), Steven Moss, Pamela Murray (Special Collections & College Archives, Lafayette College), Pete Myers, Benjamin Panciera (Linda Lear Center for Special Collections & Archives, Connecticut College), Richard L. Paul, Miriam Pawel, Dawn Peterson, Donald Pharr, Dorothy Porter, Carolina Reid, Naomi Rogers, David and Lois Roper, David W. Rose (March of Dimes Archives), Tom Rosenbaum (Rockefeller Archive Center), Caitlin Rosenthal, Dean Rowan (Berkeley Law Library), Jennifer Rusiecki, Adam Sarvana, John Saurenman, Elena Schneider, Bob Shields, Michael Sholinbeck (University of California, Berkeley Bioscience, Natural Resources, and Public Health Library), Ellen Griffith Spears, Madeleine Stebbins, Carey Stumm (National Archives at New York City), Jennifer Thomson, Paolo Toniolo, David Vail, M. Indira Venkatesan, Melissa Veronesi, James Webb, Mary Piowaty West, Christine Whelan,

Mary S. Wolff, Shalis Worthy (Huntsville–Madison County Public Library System), Charles Wurster, Victor Yannacone, and Anna Young. Extra-special thank-yous to Wendy Strothman, Hillary Brenhouse, Remy Cawley, Ben Platt, Maddie Lipscomb, Jonathan Kuo, Karen and Mike Remais, Bill and Georganne Conis, and, from the bottom of my heart, Justin, Rey, Molly, and Bailey, rest his sweet, warm soul.

SOURCE NOTES

ARCHIVES AND ABBREVIATIONS

ADAH .... Alabama Department of Archives and History

ADP .... Albert Deutsch Papers, Melvin Sabshin, M.D. Library and Archives, American Psychiatric Association

AEA .... Anti-Environmental Archives, Climate Investigations Center

AMA .... American Medical Association

BSP/Syracuse .... Benjamin Spock and Mary Morgan Papers, Syracuse University Special Collections Research Center

EDFA .... Environmental Defense Fund Archives, Stony Brook University Special Collections and Archives

EWOH/Bancroft .... Earl Warren Oral History Project, Bancroft Library, University of California, Berkeley

FMDP/UCSD .... Farmworker Movement Documentation Project, University of California, San Diego Library

FSP/Countway .... Fredrick Stare Papers, Harvard Countway Library of Medicine

GA .... Georgia Archives

HMCPL .... Huntsville–Madison County Public Library

LLC/RC .... Linda Lear Collection of Rachel Carson, Connecticut College

LOP/WHS .... Lorrie Otto Papers, Wisconsin Historical Society'

MARBL...................... Manuscript and Rare Book Library, Emory University

MOD .... March of Dimes Archive

MSFCHP .... Marshall Space Flight Center History Project

NARA Atlanta............ National Archives and Records Administration, Morrow, Georgia

NARA College Park ... National Archives and Records Administration, College Park, Maryland

NYPL......................... Manuscripts and Archives Division of the New York Public Library

RAC............................ Rockefeller Archive Collection

RCP/Beinecke............. Rachel Carson Papers, Beinecke Rare Book & Manuscript Library, Yale University

SHI............................. Science History Institute, Philadelphia

TD.............................. Toxic Docs, Columbia University and City University of New York

TJP/Bancroft.............. Thomas Jukes Papers, Bancroft Library, University of California, Berkeley

UCDRC...................... University of Cincinnati Digital Resource Commons

UCSF/IDL.................. University of California, San Francisco Industry Documents Library

UFW/WRL .... United Farm Worker Records, Walter P. Reuther Library, Detroit

UMFHP...................... University of Michigan Faculty History Project

WDP/Vanderbilt......... William Darby Papers, Eskind Biomedical Library Manuscripts Collection Repository, Vanderbilt University

WHOA........................ World Health Organization Archives, Geneva, Switzerland

INTRODUCTION

For every chapter of this book I relied on far more sources than I've been able to list here. In several cases I tracked down people with direct or inherited memories of events I chose to describe in the book, but in every case I reconstructed events from historical documents first and oral histories and personal communication second. These chapter notes highlight some of the most important sources for each chapter and provide source information for direct quotations. For further reading on actors who have manipulated public understanding of science and their tactics and motives, see, for example, Naomi Oreskes and Eric Conway's *Merchants of Doubt*, Naomi Oreskes's *Why Trust Science?*, David Michaels's *Doubt Is Their Product* and *The Triumph of Doubt*, and Allan Brandt's *Cigarette Century*. Direct quotations in the Introduction come from documents in the Robert Woodruff Papers at Emory University's Rare Manuscript and Rare Book Library and the University of California San Francisco's Industry Documents Library; Anthony Standen, "DDT," *Life*, July 8, 1946, 47–54; Allan Brandt, *The Cigarette Century* (New York: Basic Books, 2007), 204; and conversations with scientists I spoke to over the course of my research.

PROLOGUE

I first came across documents and articles describing the events in Triana, Alabama, at the National Archives in Atlanta (NARA Atlanta); research there led me to court files kept at the Alabama Department of Archives and History (ADAH) in Montgomery. A handful of scholars have written about Triana and Clyde Foster, including Dorceta Taylor (*Toxic Communities*), Bruce E. Johansen (*Environmental Racism in the United States and Canada*), and Richard Paul and Steven Moss (*We Could Not Fail: The First African Americans in the Space Program*). Paul put me in touch with Clyde Foster's daughter, Edith Foster, who lives next door to Foster's widow, Dorothy, and who graciously spent hours on the phone with me and sent me documents from her own and her father's files when the COVID-19 pandemic canceled my plans to travel to Triana. I also relied on government documents, countless press clippings, and other interviews, mentioned in the notes for chapters 20 and 22, to reconstruct the events described herein. Directly quoted material in this chapter comes from Paul and Moss, *We Could Not Fail*; Steve Hagey, "Danger Seen in Eating Fish from Rivers," *Tennessean*, December 8, 1978, 26; "DDT Legacy Troubles Triana," *Environmental Planning Impact*, February 1979, 2, 7–8; Bob Dunnavant, "Study Shows Triana Residents Ate Unsafe Fish," *Birmingham (AL) Post-Herald*, February 1, 1979, box SG018685, folder: DDT Newspaper Articles, Attorney General/Office, ADAH; Hamilton Bims, "Rocket Age Comes to Tiny Triana," *Ebony*, March 1965, 106–112; D. Michael Cheers, "Main Source of Food for Blacks in Alabama Town: Fish Poisoned with DDT," *Jet*, March 6, 1980, 14–16; Tennessee Valley Authority, "Triana—A Town with DDT in Its Blood," in *Tennessee Valley Authority Annual Report*, 1979, 28–41; and Thomas Noland, "Triana Fish Story," *Southern Changes* 1, no. 8 (1979): 14–15.

CHAPTER I

DDT's discovery has been written about in countless texts. The account in this chapter is based on writings by Victor Froelicher and his son Franz Froelicher, including materials and information provided to me by Franz Froelicher's widow, Marguerite Froelicher, and sister, Madeleine Stebbins, and a broad collection of government documents, scientific articles, and local and national news accounts. My decisions to focus on the Froelicher family and Edward Knipling were narrative choices. Many individuals in and outside of the US government played key roles in DDT's introduction and prioritization but are not mentioned in this chapter or elsewhere in this volume. For alternative accounts of DDT's discovery, see, for example, Thomas R. Dunlap, *DDT: Science, Citizens, and Public Policy* (Princeton, NJ: Princeton University Press, 1981); David Kinkela, *DDT and the American Century* (Chapel Hill: University of North Carolina Press, 2011); and Frederick Rowe Davis, *Banned: A History of Pesticides and the Science of Toxicology* (New Haven, CT: Yale University Press, 2014). Directly quoted material in this chapter comes from James C. Whorton, *Before Silent Spring* (Princeton, NJ: Princeton University Press, 1974), 182, 179; "Beetle Spray Kills Seven and Makes More Than Thousand Ill," *Philadelphia Inquirer*, August 21, 1925, 1; Robert Rice, "A Reporter at Large: DDT," *New Yorker*, July 17, 1954, 31–56; Victor Froelicher, "The Story of DDT," *Soap and Sanitary Chemicals* 20, no. 7 (July 1944): 115–118, 145; T. F. West and G. A. Campbell, *DDT: The Synthetic Insecticide* (London: Chapman & Hall, 1946), 6; Hal Higdon, "Birth Control Replaces DDT," *Tampa Tribune*, January 11, 1970, 33; A. H. Madden, Arthur W. Lindquist, and E. F. Knipling, "DDT as a Residual Spray for the Control of Bedbugs," *Journal of Economic Entomology* 37, no. 1 (February 1944): 127–128; E. F. Knipling, "The Development and Use of DDT for the Control of Mosquitoes," *Journal of the National Malaria Society* 4, no. 2 (June 1945): 77–92; Dan Chapman, "Dusting Off DDT's Image," *Atlanta Constitution*, September 9, 2001, D1, D3; David Kinkela, *DDT and the American Century* (Chapel Hill: University of North Carolina Press, 2011), 20; T. F. West and G. A. Campbell, *DDT and Newer Persistent Insecticides* (New York: Chemical Publishing Co., 1952), 10; "Notes Taken at Orlando, FL School on DDT," folder: DDT Reports vols. 1 & 2, box 3, Historical Files, RG 442, NARA Atlanta; Edmund P. Russell III, "The Strange Career of DDT: Experts, Federal Capacity, and Environmentalism in World War II," *Technology and Culture* 40, no. 4 (1999): 770–796; Committee on Medical Research of the OSRD, "Insecticides and Insect Repellents Developed for the Armed Forces at the Orlando, Fla., Laboratory, RESTRICTED REPORT," July 1, 1945, folder: DDT Reports vols. 1 & 2 (vol. 2), NARA Atlanta; "Results of Examinations of Three Men Having Relatively Long Continued Occupational Exposure to

DDT," August 1, 1944, folder: DDT Reports vols. 1 & 2, NARA Atlanta; Frank Carey, "DDT, Miracle Bug Killer, Opened New Chapter in War and Peace at Saipan," *Messenger-Inquirer* (Owensboro, KY), November 27, 1944, 3; Frank Carey, "Benefits of DDT, New Insecticide, Seen as Greater Than Penicillin," *Hutchinson (KS) News*, November 26, 1944, 10; Howard V. Smith, "DDT's Secret Told," *Miami News*, March 18, 1945, 15; Roland Nicholson, "Medical Aid Great Benefit to Soldier," *Washington Post*, April 10, 1943, B1; Anthony Standen, "DDT," *Life*, July 8, 1946, 47–54; Charles Froelicher Papers, 1945, United States Holocaust Memorial Museum, Washington, DC; "DDT Available for Civilian Use Early Next Week," *Washington Post*, July 27, 1945, 1; "Public to Receive DDT Insecticide," *New York Times*, July 27, 1945, 17; "Science: Homemade DDT," *Time*, August 6, 1945, online; "State Orders DDT Colored Distinctively for Civilian Sale," *Daily Republican* (Monongahela, PA), August 6, 1945, 1; "Public Will Soon Be Able to Buy DDT in Any Quantity," *Standard-Speaker* (Hazleton, PA), August 22, 1945, 5; and Jerry Jackson, "Lab's Orlando Days Numbered," *Orlando Sentinel*, November 22, 1992, D1–D3.

CHAPTER 2

The account in this chapter and the following chapter appears in Elena Conis, "Polio, DDT, and Disease Risk in the United States After World War II," *Environmental History* 22, no. 4 (October 2017): 696–721. I relied on hundreds of local, historical newspaper accounts and historical medical journal publications to reconstruct the use of DDT to fight polio. David Rose provided me with invaluable documents from the March of Dimes archive, which tragically closed down before I was through with this book. Directly quoted material in this chapter comes from Naomi Rogers, *Dirt and Disease: Polio Before FDR* (New Brunswick, NJ: Rutgers University Press, 1992), 18–21, 189; David M. Oshinsky, *Polio: An American Story* (Oxford University Press, 2005), 23, 72; Dawn Biehler, *Pests in the City* (Seattle: University of Washington Press, 2013), 39; Robert Ward, Joseph L. Melnick, and Dorothy M. Horstmann, "Poliomyelitis Virus in Fly-Contaminated Food Collected at an Epidemic," *Science* 101, no. 2628 (May 11, 1945): 491–493; "DDT and Poliomyelitis," *Hickory Daily Record*, May 19, 1945, 4, folder: Chemical Research—DDT, 1945–1952, series 14: Poliomyelitis, Medical Records Program, March of Dimes (MOD) Archive (all subsequent items from this collection come from this folder); "Second DDT Round Slated in Hidalgo," *Monitor* (McAllen, TX), June 26, 1945, 1; "Texas Needs Better Polio Protection," *Monitor*, September 5, 1945, 5; "Paralysis Cases Rise: Infantile Afflictions in Nation Are Ahead of Last Year," *New York Times*, 1945, 12; "No Time to Relax," *Charlotte*

(NC) Observer, June 12, 1945, 8; "Dr. Gudakunst to C. H. Crabtree," May 24, 1945, MOD; and "memorandum from Mr. Maguire to Mr. LaPorte," August 17, 1945, MOD. Roosevelt's polio diagnosis is an ongoing source of medical dispute. See Armond S. Goldman et al., "Franklin Delano Roosevelt's (FDR's) (1882–1945) 1921 Neurological Disease Revisited; The Most Likely Diagnosis Remains Guillain-Barré Syndrome," *Journal of Medical Biography* 24, no. 4 (November 2016): 452–459; John F. Ditunno Jr., Bruce E. Becker, and Gerald J. Herbison, "Franklin Delano Roosevelt: The Diagnosis of Poliomyelitis Revisited," *PM&R* 8, no. 9 (September 2016): 883–893; José Berciano, "Additional Arguments Supporting That Franklin Delano Roosevelt's Paralytic Illness Was Related to Guillain-Barré Syndrome," *Journal of Medical Biography* 26, no. 2 (May 2018): 142–143.

CHAPTER 3

Directly quoted material comes from Joseph L. Melnick et al., "Fly-Abatement Studies in Urban Poliomyelitis Epidemics During 1945," *Public Health Reports* 62, no. 25 (June 20, 1947): 901–932; "memorandum from Fred Maguire to Don Gudakunst," August 14, 1945, MOD; "Area of Polio Infested City Sprayed with DDT Kept Secret," *Augusta (GA) Chronicle*, August 21, 1945, 1; Jane Stafford, "Airplane to Spray DDT on Entire City," *Courier-Post* (Camden, NJ), August 17, 1945, 2; John R. Paul, *A History of Poliomyelitis* (New Haven, CT: Yale University Press, 1971), 296–297; "Spray Polio-Stricken City with Army DDT," *Cushing (OK) Daily Citizen*, August 20, 1945, 1; "Army Bomber Sprays 'DDT' on City in Effort to Halt Polio," *Metropolitan Pasadena (CA) Star-News*, August 20, 1945, 5; "letter from John R. Paul to D. W. Gudakunst," September 11, 1945, MOD; "memorandum from Mr. Maguire to Mr. LaPorte"; "Boston's Schools to Be Nation's First Sprayed with DDT Against Polio," *Daily Boston Globe*, September 26, 1945, 1; "Texas Outbreak 'No Mystery,' Official Blames It on Filth," *Washington Post*, May 14, 1946, 11; "San Antonio Dusting DDT Many Areas," *Waxahachie (TX) Daily Light*, May 15, 1946, 1; "Mainly About Pampa and Her Neighbor Towns," *Pampa (TX) Daily News*, July 12, 1946, 8; "Kill Polio Carrier Flies," *Corpus Christi (TX) Caller-Times*, June 16, 1946, 13; Oshinsky, *Polio*, 3; "San Antonio Move Halts Polio Spread," *Bonham (TX) Daily Favorite*, May 13, 1946, 1; Frank K. Steinrock, "DDT, TIFA, and the Polio Epidemic at San Antonio, Texas," *Mosquito News* 6, no. 3 (1946): 141–153; C.-E. A. Winslow, "Panic or Reason in Dealing with Poliomyelitis?," *American Journal of Public Health and the Nation's Health* 36, no. 8 (August 1946): 916–918; "City Passes Milk Order," *Light* (San Antonio, TX), August 5, 1946, 1; Edward Hughes, "Super

Fly, Thriving on D.D.T., Is Called No Immediate Threat," *Wall Street Journal*, January 26, 1948, 1; Associated Press, "Flies Become Immune to DDT," *Tampa Bay Times*, January 17, 1949, 11; Harry Weaver, "Draft—'Spraying of DDT,'" March 8, 1950, MOD; and Paul F. Ellis, "DDT Spray for Polio Found Ineffective," *Bend (OR) Bulletin*, July 16, 1949, 4.

CHAPTER 4

Directly quoted material in this chapter came from John Phillips Street, *The Composition of Certain Patent and Proprietary Medicines* (Chicago: AMA, 1917), 265; *The College of the City of New York Sixty-First Annual Register (1909–1910): Announcement of Courses for 1910–1911* (New York: Martin B. Brown, 1910); "Personal," *City College Quarterly*, March 1913, 161; Max Sobelman, "Statement by Max Sobelman to DDT Hearing Officer, Washington State Department of Agriculture, October 15, 1969," *DDT: Selected Statements from State of Washington DDT Hearings and Other Related Papers* (DDT Producers of the United States, 1970). Other sources that provided key background information for this chapter include David A. Hounshell and John Kenly Smith, *Science and Corporate Strategy: Du Pont R & D, 1902–1980* (Cambridge: Cambridge University Press, 1992), 453; Answer of Defendant Olin Corporation to the Amended Complaint of the United States of America, March 25, 1982, Case Number CV 80 PT 5300 NE, Civil Case Screening Project, US District Court of Alabama, Northern Division (Birmingham), Records of the District Courts of the United States, box 710, NARA Atlanta; Committee on Review of the Conduct of Operations for Remediation of Recovered Chemical Warfare Materiel from Buried Sites, Board on Army Science and Technology, and Division on Engineering and Physical Sciences, "Redstone Arsenal: A Case Study," in *Remediation of Buried Chemical Warfare Materiel* (Washington, DC: National Academies, 2012); Helen Brents Joiner, "The Redstone Arsenal Complex in the Pre-missile Era: A History of Huntsville Arsenal, Gulf Chemical Warfare Depot, and Redstone Arsenal, 1941–1949," US Army Aviation and Missile Life Cycle Management Command, 1966; Terence Kehoe and Charles David Jacobson, "Environmental Decision Making and DDT Production at Montrose Chemical Corporation of California," *Enterprise & Society* 4, no. 4 (2003): 640–675; US Tariff Commission, *Synthetic Organic Chemicals: United States Production and Sales, 1947*, vol. 162 (Washington, DC: US Government Printing Office, 1949); and items in the DDT vertical file of the Huntsville–Madison County Public Library (HMCPL). I also relied on depositions given in cases and investigations involving Montrose in New Jersey and California.

CHAPTER 5

The account in this chapter and Chapter 7 was first published, in a different form, in the digital humanities journal *Southern Spaces* as Elena Conis, "DDT Disbelievers: Health and the New Economic Poisons in Georgia After World War II," October 28, 2015, available at www.southernspaces.org and based on the letters in the folder T-47: Toxicology—Economic Poisons—Insecticides—Mrs. Plyler & Colson, in box 3, series 21, in the Georgia Archives in Atlanta, GA. Colson and Plyler's letters were quite detailed; I also relied heavily on health department files in the same series. David Roper and the *Claxton Enterprise* digital archives helped fill in gaps in the letters. Directly quoted material in this chapter comes from Mrs. H. J. Colson to the National Health Council, January 31, 1949, Georgia Archives; Mrs. Bogan C. Plyler to Prince H. Preston, September 7, 1950, Georgia Archives; "The Magic Wonder Drug DDT Not a Panacea for All Bugs," *Capper's Weekly*, October 13, 1945, 46; Channing Cope, "DDT Experiment Proves Successful," *Atlanta Constitution*, August 29, 1945, 7; Mrs. Henry J. Colson to Guy G. Lunsford, July 2, 1948, Georgia Archives; Mrs. H. J. Colson to Lester M. Petrie, June 30, 1948, Georgia Archives; W. D. Lundquist to Guy G. Lunsford, July 17, 1948, Georgia Archives; Mrs. H. J. Colson to Guy G. Lunsford, July 12, 1948, Georgia Archives; The Federal Insecticide, Rodenticide, and Fungicide Act, P.L. 80-104, 61 Stat. 163 (1947) ch. 125; "Spray Poison Kills Two Tots," March 17, 1954, and undated news clipping, "Poisonous Spray Kills Boy 6," record group 26, subgroup 4, series 21, box 3, folder: Toxicology-General, Georgia Archives; Arnold J. Lehman to L. M. Petrie, July 2, 1948, Georgia Archives; Lawrence T. Fairhall to L. M. Petrie, June 29, 1948, Georgia Archives; L. M. Petrie to Guy G. Lunsford, August 10, 1948, Georgia Archives; Mrs. B. C. Plyler to L. M. Petrie, December 14, 1950, Georgia Archives; L. M. Petrie to Mrs. H. J. Colson, February 4, 1949, Georgia Archives; Mrs. H. J. Colson to L. M. Petrie, January 29, 1949, Georgia Archives; Jack Harrison Pollack, "The Shame of Our Local Health Departments," *Collier's Weekly*, January 22, 1949, 13; Mrs. H. J. Colson to the National Health Council, January 31, 1949, Georgia Archives; and newspaper clipping, "Milk Poisoned by DDT Spray, Farmers Warned," *Atlanta Journal*, March 28, 1949, Georgia Archives.

CHAPTER 6

I was surprised to find so little written about Morton Biskind or Virus X. Most of this chapter is based on Biskind's testimony, his published scientific articles, newspaper articles that mentioned him, and letters that he exchanged with Rachel Carson, held in archival collections at Yale and the University of Connecticut. Information on Virus X comes entirely from scientific journals and newspaper articles. Directly quoted material comes from "'Virus X' Termed Influenza Centered in Intestinal Tract," *Palladium-Item* (Richmond, IN), January 25, 1948, 7; "'Virus X' Is Suspected in Wave of Minor Ills," *Miami News*, December 30, 1948, 17; "Virus X Made Up of Two Different Common Illnesses," *Palladium-Item*, December 31, 1947, 4; Morton S. Biskind, "DDT Poisoning and the Elusive 'Virus X': A New Cause for Gastro-Enteritis," *American Journal of Digestive Diseases* 16, no. 3 (March 1949): 79–84; R. A. M. Case, "Toxic Effects of D.D.T. in Man," *British Medical Journal* 2, no. 4432 (December 15, 1945): 842–845; Biskind, "DDT Poisoning and X Disease in Cattle," *Journal of the American Veterinary Medical Association* 114 (January 1949): 20; "X Disease in Cattle," *Prairie Farmer*, August 28, 1948, 31; Biskind, "Special Notice! DDT Poisoning a Serious Public Health Hazard," *American Journal of Digestive Diseases* 16, no. 2 (February 1949): 73; Albert Deutsch, "DDT—Is It Most Deadly Poison Ever Unleashed on Humanity?," *York (PA) Daily Record*, April 1, 1949, 1; Albert Deutsch, "DDT and You," *New York Post*, April 1, 1949, 1; staff correspondent, "Report That DDT Is 'Mystery Poison' Causes Concern in U.S.; Health Chiefs Confer," *Sydney (NSW) Morning Herald*, n.d., 1; Albert Deutsch, "DDT and You! It Can Cost You Your Health," *New York Post*, March 30, 1949, 2; N. S. Haseltine, "U.S. Chemists Checking Milk Shipments for Any DDT Spray: Milk Checked for DDT Trace," *Washington Post*, April 1, 1949, 1; and "Says Congress Should Check on Food Chemicals," *Nashua (NH) Telegraph*, May 9, 1949, 11.

CHAPTER 7

See my note for Chapter 5, above. Directly quoted material comes from Federal Security Agency, Public Health Service, Washington, DC Memorandum, April 1, 1949, record group 26, subgroup 4, series 21, box 3, folder T-47: Toxicology—Economic Poisons—Insecticides—Mrs. Plyler & Colson, Georgia Archives; L. M. Petrie to W. D. Lundquist, July 1, 1949, Georgia Archives; S. W. Simmons to L. M. Petrie, August 22, 1949, Georgia Archives; Justus C. Ward to L. M. Petrie, August 19, 1949, Georgia Archives; *Warning Labels: A Guide for the Preparation of Warning Labels for Hazardous Chemicals*, Manual L-1 (Washington, DC: Manufacturing Chemists' Association, 1949); newspaper clipping, "Economic Poisons Bulletin," November 1950, record group 26, subgroup 4, series 21, box 3, folder N-16: Newspaper Clippings—Insecticides, Georgia Archives; Mrs. H. J. Colson to James G. Townsend, August 20, 1949, Georgia Archives; Mrs. H. J. Colson to L. M. Petrie, August 14, 1949, Georgia Archives; Mrs. B. C. Plyler to Hon. Herman Talmadge, August 17, 1950, Georgia Archives; Executive Department, State of Georgia to

Mrs. B. C. Plyler, August 25, 1950, Georgia Archives; Mrs. B. C. Plyler to L. M. Petrie, December 4, 1950, Georgia Archives; Channing Cope, "DDT Experiment Proves Successful," *Atlanta Constitution*, August 29, 1945, 7; Reverend W. M. Bishop to Governor Herman Talmadge, September 15, 1950, Georgia Archives; Economic Poisons Study, October 11, 1950, Georgia Archives; W. D. Lundquist, "Trip Report," October 11, 1950, Georgia Archives; Petition to T. F. Sellers, October 24, 1950, Georgia Archives; Mrs. B. C. Plyler to Richard H. Fetz, April 27, 1951, Georgia Archives; R. H. Fetz to Mrs. Anne E. Brewton, November 14, 1951, Georgia Archives; and "Spray Program Started to Reduce Pest Threat," *Claxton Enterprise*, August 17, 1952, 1.

CHAPTER 8

See notes for Chapter 6, above. Directly quoted material comes from *Chemicals in Food Products: Hearings Before the House Select Comm. to Investigate the Use of Chemicals in Food Products*, 81st Cong. (1950), 335, 2, 31, 32, 34, 701, 712, 705, 721, 706, 707; *Chemicals in Food Products: Hearings Before the House Select Comm. to Investigate the Use of Chemicals in Food Products*, 82nd Cong. (1950), 161, 90, 94, 95, 97, 99, 110; "Food Chemicals Seen Raising Big Problem," *New York Times*, January 4, 1951, 47; Paul P. Kennedy, "Food Firms Balky, House Group Finds," *New York Times*, December 3, 1950, 11; Dan Markel, "Russ Blamed for Epidemics in U.S. by Intelligence Aid," *San Francisco Examiner*, October 17, 1948, 17; Small and Hadley, "Leukocytes in Virus X Infection"; and "West Coast Doctor Identifies Virus X," *Union City (IN) Times-Gazette*, December 31, 1947, 2. Note: although Dunbar was scheduled to be the first witness, the committee allowed University of Chicago physician Anton Julius Carlson to testify before him because of a scheduling conflict.

CHAPTER 9

Directly quoted material for this chapter comes from *Chemicals in Food Products*, 82nd Cong. (195), 400, 391, 386, 149, 937, 150; Davis, *Banned*, 144; Select Comm. to Investigate the Use of Chemicals in Food Products, H.R. Rep. No. 82-2356 (1952), 3, 4, 41; Minutes of the Fifteenth Meeting of the Directors of the Manufacturing Chemists' Association, January 8, 1952; "Chemicals in Food," *Mexico (MO) Ledger*, March 4, 1952, 10; 68 Stat. 511 1954; Perkins, *Insects, Experts, and the Insecticide Crisis*, 30; the Pesticide Residues Amendment of 1954, Pub. L. No. 83-518, ch. 559, 68 Stat. 511; Nancy Langston, *Toxic Bodies: Hormone Disruptors and the Legacy of DES* (New Haven, CT: Yale University Press, 2010), 81; Wayland J. Hayes Jr., William F. Durham, and Cipriano Cueto Jr., "The Effect of Known Repeated

Oral Doses of Chlorphenothane (DDT) in Man," *Journal of the American Medical Association* 162, no. 9 (October 27, 1956): 890–897; Wayland J. Hayes Jr., William E. Dale, and Carl I. Pirkle, "Evidence of Safety of Long-Term, High, Oral Doses of DDT for Man," *Archives of Environmental Health* 22, no. 1 (January 1971): 119–135; and Richard E. L. Fowler, "Insecticide Toxicology, Manifestations of Cottonfield Insecticides in the Mississippi Delta," *Journal of Agricultural and Food Chemistry* 1, no. 6 (June 1953): 469–473. Note: historian Sarah Vogel says that the 1958 law shifted regulation of chemicals in food from the "per se rule" to the "de minimus" standard—chemical risks were dependent on the size of the exposure, not on the presence of a hazardous substance. In fact, the Miller amendment had already achieved this. Sarah A. Vogel, "Forum: From 'The Dose Makes the Poison' to 'The Timing Makes the Poison,'" *Environmental History* 13 (October 2008): 667–673.

CHAPTER 10

All directly quoted material in this chapter comes from the Paul F. Russell Diaries, 1950–1958, Rockefeller Archive Center (RAC), Sleepy Hollow, New York, and selected documents from the World Health Organization Archives in Geneva, Switzerland, thanks to the help of Reynald Erard. The political cartoon described in the chapter is reproduced in Kinkela, *DDT and the American Century*, 52.

CHAPTER 11

This chapter is based on letters that Marjorie Spock exchanged with Rachel Carson, her courtroom testimony, and other writings and accounts of her life and work. Directly quoted material comes from court transcript, Murphy et al. v. Benson, February 10, 1958, folder Murphy et al. v. Benson pp. 1472–1609, box B-38, National Audubon Society Records, Manuscripts and Archives Division, the New York Public Library (NYPL), Astor, Lenox, and Tilden Foundations, 8–9, 22, 39, 1483; Marjorie Spock, "DDT Trial News," Letters, *Organic Gardening and Farming*, March 1958; Henry Barnes, *Into the Heart's Land: A Century of Rudolf Steiner's Work in North America* (Great Barrington, MA: Steiner Books, 2005); Linda Lear, *Rachel Carson: Witness for Nature* (Boston: Mariner, 1997), 318, 327; Ehrenfried E. Pfeiffer, "Do We Really Know What We Are Doing? DDT Spray Programs—Their Value and Dangers," *Biodynamics Journal* 45 (1958), 2–40; John Paull, "The Rachel Carson Letters and the Making of *Silent Spring*," *SAGE Open* 3, no. 3 (July 2013), 1–12; Marjorie Spock to Mr. Stuart Huckins, February 15, 1958, folder 856, box 45, Rachel Carson Papers, series I, Writings, box 45, folder

855: DDT Controversy, Yale University Beinecke Rare Book & Manuscript Library, hereafter RCP/Beinecke; trial transcript, box 45, folder 855, RCP/ Beinecke; Robert Cushman MURPHY et al., and Archibald B. Roosevelt et al., Plaintiffs, v. Ezra Taft BENSON, etc., et al., Defendants, US District Court for the Eastern District of New York: Murphy v. Benson, 164 F. Supp. 120 (D.N.Y. 1958); William Longgood, "Pesticides Poison Us," *American Mercury*, July 1958, 33–54; Marjorie Spock, notes, March 10, 1958, folder 856, box 45, RCP/Beinecke; and Mark Hamilton Lytle, *The Gentle Subversive: Rachel Carson, Silent Spring, and the Rise of the Environmental Movement* (New York: Oxford University Press, 2007), 2, 133. Notes: Carson's biographer, Linda Lear, found no indication that Carson was aware of White's early article and suspects that she contacted the writer in order to reach his influential editor. This series of events is described in greater detail in Linda Lear, *Rachel Carson: Witness for Nature* (Boston: Mariner, 1997), 312–322. Regarding cellular oxidation, Pfeiffer brought up this subject first. He pointed out that DDT inhibited enzyme reactions necessary for cell oxidation as well as the cytochrome C oxidase enzyme.

CHAPTER 12

For this chapter, I became indebted to the enormous literature on Rachel Carson and *Silent Spring*. For just a few examples, see Maril Hazlitt, "Voices from the Spring: *Silent Spring* and the Ecological Turn in American Health," in *Seeing Nature Through Gender*, ed. Virginia Scharff (Lawrence: University Press of Kansas, 2003), 103–128; Priscilla Coit Murphy, *What a Book Can Do* (Amherst: University of Massachusetts Press, 2007); *On a Farther Shore: The Life and Legacy of Rachel Carson, Author of* Silent Spring (New York: Broadway, 2013); Paul Brooks, *The House of Life: Rachel Carson at Work* (Boston: G.K. Hall, 1985); Christopher Sellers, *Crabgrass Crucible: Suburban Nature and the Rise of Environmentalism in Twentieth-Century America* (Chapel Hill: University of North Carolina Press, 2012); Lear, *Rachel Carson: Witness for Nature*, 1997; Lytle, *The Gentle Subversive*, 2007. Directly quoted material comes from Murphy v. Benson, 164 F. Supp. 120 (E.D.N.Y. 1958); Marjorie Spock to Mr. Huckins, February 15, 1958, folder 856, box 45, RCP/Beinecke; Paull, "The Rachel Carson Letters and the Making of *Silent Spring*"; Spock, "Portrait of Rachel"; Rachel Carson to Marjorie Spock, October 12, 1959, RCP/Beinecke; Rachel Carson to Edwin Way Teale, cited in Lear, 334; Rachel Carson to Morton Biskind, December 3, 1959, folder 762, box 42, RCP/Beinecke; Rachel Carson to Marjorie Spock, January 18, 1960, folder 829, box 44, RCP/Beinecke; Rachel Carson to Marjorie Spock, July 11, 1960, folder 828, box 44, RCP/Beinecke;

Lear, 365; William J. Darby, "The Poisons in Your Food," *Science* 131, no. 3405 (April 1960): 979; Robert Cushman Murphy et al., petitioners, v. Lloyd Butler, Area Supervisor, etc., et al., 360 U.S. 929, 750 (1960) (Douglas dissenting opinion); Barnes, *Into the Heart's Land,* 2005; Rachel Carson to Marjorie Spock, April 12, 1960, folder 829, box 44, RCP/Beinecke; Morton Biskind to Rachel Carson, February 2, 1962, folder 762, box 42, RCP/Beinecke; Rachel Carson, *Silent Spring* (New York: Houghton Mifflin, 2002), 238, 22, 222; William J. Darby, "Silence, Miss Carson," *Chemical & Engineering News Archive* 40, no. 40 (October 1962): 60–63; and Jonathan Leonard, "The Public and *Silent Spring,*" remarks presented before the 31st Annual Meeting of the National Agricultural Chemicals Association at the Greenbrier in White Sulphur Springs, West Virginia, September 9, 1964, box 83, folder 1447, RCP/Beinecke.

CHAPTER 13

The digital collections maintained by the University of California, San Diego Library were a gold mine for this chapter. Background material for the chapter came from Frank Bardacke, *Trampling Out the Vintage: Cesar Chavez and the Two Souls of the United Farm Workers* (New York: Verso, 2011), 66–68, 84; Ruben Garcia, *Marginal Workers: How Legal Fault Lines Divide Workers and Leave Them Without Protection* (New York: New York University Press, 2012); Miriam Pawel, *The Crusades of Cesar Chavez: A Biography* (New York: Bloomsbury, 2014), 149; and Randy Shaw, *Beyond the Fields: Cesar Chavez, the UFW, and the Struggle for Justice in the 21st Century* (Berkeley: University of California Press, 2008), 1, 18. Directly quoted material comes from Lori Flores, *Grounds for Dreaming* (New Haven, CT: Yale University Press, 2016), 46; Thomas H. Milby and Fred Ottoboni, "Report of an Epidemic of Organic Phosphate Poisoning in Peach Pickers, Stanislaus County, California" (Bureau of Occupational Health, Department of Public Health, State of California, December 23, 1963); Phillip Hoose, *We Were There, Too! Young People in U.S. History* (New York: Farrar, Straus and Giroux, 2001), 230; Jack Ostrich, "Irma West—Memories of an Early Woman Doctor," *Sierra Sacramento Valley Medicine,* October 2015, 12–15; Irma West, "Control of Insecticide Exposure in Employment—A Guide to Physicians for Dealing with Organic Phosphates," *California Medicine* 86, no. 5 (May 1957): 325–330; Nils P. Larsen, "The Men with Deadly Dreams," *Saturday Evening Post,* December 3, 1955, 20–21, 140–143; "Braceros' Deaths Pose Big Puzzle," *Press-Telegram* (Long Beach, CA), August 8, 1957, B7; Carson, *Silent Spring,* 30; *United States Congress, Congressional Record: Proceedings and Debates of the 87th Congress Second Session,* vol. 108 (US Government Printing Office, 1962), 17647; Lawrence E. Davies, "Coast Dairymen Battle Pesticides," *New York Times*, August 25, 1962, 44; and *Pesticides and Other Agricultural Chemicals as a Public Health Problem in California: Hearings Before the Subcomm. on Reorganization and International Organizations of the Comm. on Government Operations*, 88th Cong. 636–640 (1963) (statement of Irma West, Medical Officer, Bureau of Occupational Health, California State Department of Public Health). Sadly, I was unsuccessful in my efforts to reach Govea's family members, but I was delighted and fortunate to track down Irma West's son and daughter-in-law, who shared stories and personal papers in their family's possession.

CHAPTER 14

Material for this chapter came primarily from the University of California, San Diego Library Farmworker Movement Documentation Project, the UFW Records at the Walter P. Reuther Library in Detroit, and Irma West's papers and publications. Directly quoted material comes from Frank L. Campbell to William J. Darby, 29 January 1964, series 1, box 11, William Darby Papers, Eskind Biomedical Library Manuscripts Collection Repository, Vanderbilt University; René Dubos, "Environmental Biology," *BioScience* 14, no. 1 (1964): 11–14; A. A. Rosen and H. F. Kraybill, "Preface, Introduction," in *Organic Pesticides in the Environment*, Advances in Chemistry 60 (American Chemical Society, 1966), vii–x; Irma West, "Biological Effects of Pesticides in the Environment," in *Organic Pesticides in the Environment*, Advances in Chemistry 60 (American Chemical Society, 1966), 38–53; Chavez, Rodriguez, and Govea, "Richard Chavez—Marta Rodriguez—Jessica Govea," interview, Jessica Govea Oral History, Farmworker Documentation Project, University of California, San Diego Library (FMDP/UCSD); Spiros De Bono, "Labor," *Windsor (ON) Star*, November 7, 1968, 44; Cesar E. Chavez and Pete G. Velasco to Irma M. West, June 23, 1971, collection of Mary West Piowaty; "United Farm Workers Map Grape Boycott Plans, Pesticide Battle," *Sacramento Bee*, January 26, 1969, A7; "The Grape Boycott Goes On!," editorial, *El Malcriado: The Voice of the Farm Worker* III, no. 6 (June 1969): 4; United Farm Workers Organizing Committee, "Pesticides: The Poisons We Eat," 1969, FMDP/UCSD; Dolores Huerta and Jerry Cohen, "Statement of United Farm Workers Organizing Committee, AFL-CIO," unpublished manuscript, November 21, 1969; Rey Huerta, "Rey Huerta 1968–1975: The Most Memorable Times of Our Lives," unpublished manuscript; "Data on DDT and Parathion," *El Malcriado: The Voice of the Farm Worker* III, no. 7 (July 1969): 7, 11; "Perelli Minetti Signs Contract with UFWOC: 'Grapes of

Wrath' Turn to Grapes of Wine," *El Malcriado: The Voice of the Farm Worker* III, no. 12 (September/October 1969): 8–9; Pulido and Peña, "Environmentalism and Positionality"; Pawel, *The Crusades of Cesar Chavez*, 205; "Huelga Women and the Boycott: From the Fields to the Picket Line," *El Malcriado: The Voice of the Farm Worker* IV, no. 6 (August 1970): 12–15; and "Reflections on the Poisoning of Food and Man," editorial, *El Malcriado: The Voice of the Farm Worker* III, no. 14 (October 1969): 4.

CHAPTER 15

My research for this chapter drew heavily on the Environmental Defense Fund Archive at Stony Brook University and the Lorrie Otto Papers at the Wisconsin Historical Society. Directly quoted material comes from "Resource Group Holds Meeting," *Wausau (WI) Daily Herald*, July 19, 1966, 7; Dan Satran, "Worry Shown on DDT Effects," *Capital Times* (Madison, WI), July 4, 1966, 35; Robert L. Rudd, *Pesticides and the Living Landscape* (Madison: University of Wisconsin Press, 1964), 249; Dave Vogel, "War Casualties," *Historic Wauwatosa*, May 2015, 1, 3; "EDF Has Consumed Me," undated document, folder 3, box 7, mss 1050, Lorrie Otto Papers, Wisconsin Historical Society (LOP/WHS); Charles Wurster, *The DDT Wars* (New York: Oxford University Press, 2015), 16, 22, 40–41; "Draft of Acorn Days," record group 1, subgroup IV, series 7, sub-series 1, box 3, Environmental Defense Fund Archives, Stony Brook University Special Collections and Archives (EDFA); Lorrie Otto to the *New Yorker*, February 8, 1969, folder 3, box 7, mss 1050, LOP/WHS; Luther J. Carter, "DDT: The Critics Attempt to Ban Its Use in Wisconsin," *Science* 163, no. 3867 (February 7, 1969): 548–551; Luther J. Carter, "Environmental Defense Fund: Yannacone out as Ringmaster," *Science* 166, no. 3913 (December 26, 1969): 1603; Richard Jaeger, "DDT Bugs Students, Too, They Tell Officials," *Wisconsin State Journal*, December 13, 1968, 1; "Protest Against DDT," *Oshkosh (WI) Northwestern*, December 13, 1968, 14; Lorrie Otto to *Bioscience*, November 23, 1969, folder 3, box 7, mss 1050, LOP/WHS; Charles Wurster to Lorrie Otto, November 12, 1968, folder 3, box 7, mss 1050, LOP/WHS; and Bill Berry, *Banning DDT: How Citizen Activists in Wisconsin Led the Way* (Madison: Wisconsin Historical Society Press, 2014), 181, 60. Note: Wurster describes the exchange as involving an invitation from Otto, but Otto's personal papers from the time suggest that he persuaded her to sue and that she was readily convinced. I'm grateful to Wurster and to Victor Yannacone for sharing their memories of this time, and to Yannacone for the many historical documents he has made available on his personal website.

CHAPTER 16

Directly quoted material comes from the US National Library of Medicine's online collection, "Reports of the Surgeon General," https://profiles.nlm.nih .gov/spotlight/nn; Mrs. B. C. Plyler to Research Division of P. Lorillard Company, September 2, 1965, Lorillard Records, UCSF/IDL; letter from C. O. Jensen to H. B. Parmele, September 9, 1965, Lorillard Records, UCSF/IDL; Allan M. Brandt, *The Cigarette Century* (New York: Basic Books, 2007), 171; Griffith E. Quinby et al., "DDT Storage in the US Population," *JAMA* 191, no. 3 (January 18, 1965): 175; Chuck to C. O. Jensen, August 17, 1964, Lorillard Records, UCSF/IDL; Minutes of Tobacco Pesticide Committee, First Meeting, August 12, 1968, box 136, Liggett & Myers Records, UCSF/IDL; G. H. Geffert to H. B. Grice, Esq., and the Tobacco Research Council, December 17, 1968, box 995, British American Tobacco Records, UCSF/IDL; W. C. Monk to Thomas T. Irvin, March 4, 1970, record group 13, subgroup 1, series 2, Agriculture—Commissioner's Office, folder: Pesticides, box 140, Commissioner's Subject Files, Georgia Archives; FTR, FABRIQUES DE TABAC REUNIES S.A., HAUSERMANN, M. PESTICIDE RESIDUES IN SWISS CIGARETTES, September 14, 1871, Philip Morris Records, UCSF/IDL; and H. H. Schedel to R. B. Seidensticker, translation of the article in *Helsingin Sanomat*, August 3, 1971, box 10306, Phillip Morris Records, UCSF/IDL.

CHAPTER 17

Directly quoted material comes from Charles Wurster, "The Charge Is Biocide: DDT Stands Trial," *Audubon*, September 1969: 128–135; Richard D. James, "Periled Pesticide: Wisconsin Hearing on Bid to Ban DDT Could Affect Future of All Such Products," *Wall Street Journal*, March 4, 1969, 36; Elmer P. Wheeler to R. E. Keller, W. R. Richard, H. S. Bergen, J. E. Springate, and C. Paton, "DDT Briefing Paper," March 31, 1969, Monsanto archives, Toxic Docs (TD); Tom C. Ford to Fred J. Gehrung, R. J. Kotnour, and P. R. Wilkins, May 9, 1969, Monsanto archives, Toxic Docs; Ray Pagel, "Toxicologist Insists DDT Safe," *Green Bay (WI) Press-Gazette*, May 2, 1969, A3; Thomas H. Jukes, "Its Use Saves Untold Lives," *Washington Post*, May 4, 1969, B1; poem, undated, folder 1, box 1, Thomas Jukes Papers, Bancroft Library, University of California, Berkeley (TJP/Bancroft); Thomas Jukes to Philip Abelson, editor of *Science*, July 11, 1969, box 1, folder 33, TJP/Bancroft; Thomas Jukes, "Statement by Thomas Jukes at Seattle Hearings on DDT," in *Selected Statements from State of Washington DDT Hearings Held in Seattle, October 14, 15, 16, 1969, and Other Related Papers*, compiler Max Sobelman

(Diamond Shamrock Corporation, Lebanon Chemical Corporation, Montrose Chemical Corporation of California, and Olin Corporation, 1970), 3; Charles Wurster, "DDT Goes to Trial in Madison," *BioScience* 19, no. 9 (September 1969): 809–813; letter from Thomas Jukes to Alec Nisbett, BBC-TV, June 25, 1975, folder 24, box 1, TJP/Bancroft; Charles Wurster, "DDT and Robins," *Science* 156 (1968): 1413–1414; Robert C. Bjorklund, "Attorney for Group Seeking Ban on DDT Removed from Post," *Wisconsin State Journal* (Madison, WI), January 9, 1970, sec. 2, 3; Dunlap, *DDT*, 208, 228; Brief for Petitioners, Env't Def. Fund, Inc. v. Hardin, 428 F.2d 1093 (D.C. Cir. 1970) (No. 23813); Environmental Defense Fund, "Is Mother's Milk Fit for Human Consumption?," *New York Times*, March 29, 1970, 6E; Adam Rome, *The Genius of Earth Day: How a 1970 Teach-In Unexpectedly Made the First Green Generation* (New York: Hill and Wang, 2013), 196; Gary Blonston, "Anti-pollution Heroes Emerge," *Detroit Free Press*, March 11, 1970, A1; Charles F. Wurster, Affidavit, EDF v. EPA, January 22, 1971, record group 1, subgroup IV, series 7, sub-series 1, box 4, EDFA; Paul F. Russell, letter, *Science* 177, no. 4047, pp. 387–388; William Ruckelshaus, Opinion of the Administrator, In re: Consolidated DDT Hearings, Before the Environmental Protection Agency, June 2, 1972; and Federal Pesticide Control Act of 1971: Hearings Before the H. Comm. on Agriculture, 92nd Cong. 266–68 (1971) (statement of Edward Lee Rogers, general counsel, the Environmental Defense Fund).

CHAPTER 18

Material for this chapter came from Jack Hartsfield's reporting for the *Huntsville Times*, newspaper reports on Olin and the Alabama fish kills, the EDF archives, and the Triana sources described above in the notes for the Prologue.

CHAPTER 19

Directly quoted material in this chapter comes from Public Law 86-139, "An Act to Amend the Federal Insecticide, Fungicide and Rodenticide Act," August 7, 1959, https://uscode.house.gov/statutes/pl/86/139.pdf; Charles F. Wurster, Affidavit, EDF v. EPA, 22 January 1971, record group 1, subgroup IV, series 7, sub-series 1, box 4, EDFA; Dunlap, *DDT*, 228; William Ruckelshaus, Opinion of the Administrator, In re: Consolidated DDT Hearings, Before the Environmental Protection Agency, June 2, 1972; and *Federal Pesticide Control Act of 1971: Hearings Before the H. Comm. on Agriculture*, 92nd Cong. 266–268 (1971) (statement of Edward Lee Rogers, general counsel, the Environmental Defense Fund).

CHAPTER 20

The events in this chapter and the following chapter are based largely on articles and other documents held at the Alabama Department of Archives and History and the National Archives at Atlanta. Neither Foster nor Hayes was still alive when I began my research, unfortunately, but I was able to find Kathleen Kreiss, who answered my questions and shared documents of her own, and Tom Mason, who also generously shared documents and memories. Directly quoted material in this chapter comes from Tennessee Valley Authority, Water Quality and Ecology Branch, "Where the Water Isn't Clean Anymore: A Survey of Water Quality in the Tennessee Valley" (Chattanooga, 1978); D. Michael Cheers, "Main Source of Food for Blacks in Alabama Town: Fish Poisoned with DDT," *Jet*, March 6, 1980, 14–16; Bob Dunnavant, "Triana—After DDT? Another Triple-Letter Threat Lurks," *Birmingham (AL) Post-Herald*, June 11, 1980, F1; and Bob Dunnavant, "Waste Lagoon May Be Source of Triana Chemical Pollution," *Birmingham (AL) Post-Herald*, June 6, 1980, A6. Accounts of the march differ; a short article in the *Birmingham News* said that ninety people marched. Pat Houtz, "No Unusually High Levels of PCBs Found in Fish Taken from Tennessee River," *Birmingham (AL) News*, June 15, 1980, AAGO/ADAH.

CHAPTER 21

Edith Foster was always ready to answer my questions about her father, look at photos I had found, share photos of her own, or relay my questions to her mother, Dorothy. Other directly quoted material in this chapter comes from United Press International, "Corps Proposes $88 Million DDT Cleanup," *Birmingham (AL) Post-Herald*, December 30, 1980, box SG018685, AAGO/ADAH; Mike Hollis, "Foster Turns Down New Triana Tests," April 28, 1981, vertical files: DDT, HMPCL; Mary Battiata, "Awards & Asides," *Washington Post*, March 26, 1981; Joey Ledford, "DDT Spells Woe for Triana," *Tennessean* (Nashville, TN), September 8, 1981, 31; Charles Madigan, "Town Fighting Uphill Battle Against DDT," *Chicago Tribune*, November 29, 1981, 14; court transcript filed July 24, 1981, Case Number CV-79-PT-5128 NE, Civil Case Screening Project, US District Court of Alabama, Northern Division (Birmingham), Records of the District Courts of the United States, record group 21, box 720, folder vol. 4, NARA Atlanta; Affidavit of Wayland J. Hayes, Jr., June 7, 1982, Case CV 80 PT 5300 NE, box 710, Part 1 Civil Case Screening Project, US District Court of Alabama, Northern Division (Birmingham), Records of the District Courts of the United States, NARA Atlanta; and court transcript, filed June 28, 1983, James Cloud et al., plaintiffs,

v. Olin Corporation, vol. II, CASE CV 79 PT 5128 NE, Civil Case Screening Project, US District Court of Alabama, Northern Division (Birmingham), Records of the District Courts of the United States, box 721, NARA Atlanta.

CHAPTER 22

My thanks to Christine Whelan, who spent a chaotic quarantine day on the phone with me, generously sharing memories about her mother; Tom Mason, who called me right away when Marvelene Freeman gave him my number; and Bob Shields, who helped me make sense of material that I had found in the archives. Other directly quoted material in this chapter comes from Elizabeth M. Whelan, "American Council on Science and Health Third Annual Report," November 1, 1979, to October 31, 1980, Center for Science in the Public Interest, UCSF/IDL; "'At the Drop of a Rat,'" *Record* (Hackensack, NJ), April 6, 1982, B26; Hoyt Harwell, "Contaminated Fish Stir Up a Poor Town," *Los Angeles Times*, January 16, 1983, sec. Part I, 18; Mike Hollis, "Triana DDT Suits Settled," *Huntsville (AL) Times*, December 25, 1982, A1; PR Newswire, Olin Corporation Press Release, Stamford, CT, January 27, 1983; Art Harris, "Checks Came in the Mail, but the Poison Is Still in the Catfish," *Washington Post*, June 4, 1983, online; Report of Guardian ad Litem, U.S. v. Olin Corp., Case Number CV-80-PT-5300-NE, Civil Case Files, US District Court of Alabama, Northern Division (Birmingham), Records of the District Courts of the United States, box 721, NARA Atlanta; Tracy Thompson, "Problems Still Beset Triana After DDT Settlement," *Atlanta Journal/Atlanta Constitution*, July 15, 1984, A17; Tom Mason, conversation with author, May 24, 2021; Case Number CV-79-PT-5128 NE, Civil Case Screening Project, US District Court of Alabama, Northern Division (Birmingham), Records of the District Courts of the United States, record group 21, box 720, folder vol. 4, NARA Atlanta; John Peck, "DDT Plaintiffs Could Get a $22,500 Check or a 'Good News' Note," *Huntsville (AL) Times*, December 18, 1988, 1A, vertical files: DDT, Huntsville-Madison County Public Library; Francis E. McGovern, "Alabama DDT Settlement Fund," *Law and Contemporary Problems* 53, no. 4 (Autumn 1990): 61–78; and Peter Jennings and Jay Schadler, *ABC World News Tonight*, August 7, 1984, transcript.

CHAPTER 23

My deepest thanks to Pete Myers and Dianne Dumanoski, who shared more fascinating stories about the lead-up, publication, and aftermath of *Our Stolen Future* than I knew what to do with. Additional directly quoted material in this chapter comes from Theo Colborn, draft of oral history by Jody Roberts and Elizabeth McDonnell, Paonia, Colorado, August 7, 2009, Science History Institute; Theo Colborn and Coralie Clement, *Chemically-Induced Alterations in Sexual and Functional Development* (Princeton, NJ: Princeton Scientific, 1992); and Bryan Marquard, "Former Globe Publisher William Taylor II Dead at 78," *Boston Globe*, May 2, 2011.

CHAPTER 24

Mary Wolff and Barbara Cohn were invaluable sources of information about this time and so helpful in pointing me to other sources. Paolo Toniolo was gracious in responding to emailed questions from afar. Allan Chartrand, Sean Hecht, Veronika Kivenson, M. Indira Venkatesan, John Lyons, and John Saurenman were indispensable sources who shared memories, papers, and tips on what to read and track down. Directly quoted material in this chapter also comes from Paul Brodeur, "Casualties of the Workplace: Some Nonserious Violations," *New Yorker*, October 29, 1973, 44–106; M. S. Wolff et al., "Blood Levels of Organochlorine Residues and Risk of Breast Cancer," *Journal of the National Cancer Institute* 85, no. 8 (April 21, 1993): 648–652; Beth Whitehouse and Michelle Slatalla, "Finding a Voice," *Newsday* (Melville, NY), October 4, 1993, Suffolk edition, 23; John Roberts and Bob Arnot, "Link Discovered Between Breast Cancer and DDT," April 21, 1993, *CBS Morning News*, transcript; and Associated Press, "Study Links DDT to Breast Cancer," *Bradenton (FL) Herald*, April 21, 1993, A4; United Press International (UPI), "Banned Pesticide Is Safe, Scientists Claim," May 21, 1992; and D. H. Garabrant et al., "DDT and Related Compounds and Risk of Pancreatic Cancer," *Journal of the National Cancer Institute* 84, no. 10 (May 20, 1992): 764–771.

CHAPTER 25

Directly quoted material in this chapter comes from William K. Stevens, "Pesticides May Leave Legacy of Hormonal Chaos," *New York Times*, August 23, 1994, C1; Theo Colborn, draft of oral history by Jody Roberts and Elizabeth McDonnell, Paonia, Colorado, August 7, 2009, Science History Institute, 52–53; Erika Duncan, "Fostering Clean Air Through Environmental Law," *New York Times*, May 14, 1995, 13; H. Burlington and V. F. Lindeman, "Effect of DDT on Testes and Secondary Sex Characters of White Leghorn Cockerels," *Experimental Biology and Medicine* 74, no. 1 (May 1950): 48–51; Theo Colborn, Dianne Dumanoski, and John Peterson Myers, *Our Stolen Future* (New York: Plume, 1996), 198, 68, 47, vii; Theo Colborn, interview by Doug

Hamilton, *Frontline*, PBS, February 1998; Elizabeth F. Watson, "Attachment 3. Endocrine Issues Coalition Research Summary," April 1, 1996, RJ Reynolds Records, Master Settlement Agreement; and Roger Bate, "Are We Only Half the Men We Used to Be?" *Wall Street Journal*, May 7, 1996, British American Tobacco Records, UCSF/IDL.

CHAPTER 26

Jim Aidala handed me the idea for this chapter before I even realized I was looking for it. Directly quoted material in this chapter also comes from 103 Cong. Rec. 27,246 (1994); Elizabeth M. Whelan, "Repeal of Anti-Pesticide Law Is Need Whose Time Has Come," *Deseret News*, March 19, 1995, online; Howard Kurtz, "Dr. Whelan's Media Operation," *Columbia Journalism Review* 28, no. 6 (March 1990): 43–47; Larry King, "Environmental Causes of Cancer," CNN, October 27, 1994; Keith Schneider, "Ideas & Trends: A Trace of Pesticide, an Accepted Risk," *New York Times*, February 7, 1993, E6; Keith Schneider, "E.P.A. Plans to Seek Loosening of a Law on Food Pesticides: E.P.A. Is to Seek Eased Food Rule," *New York Times*, February 2, 1993, A1; John H. Cushman, "E.P.A. Settles Suit and Agrees to Move Against 36 Pesticides," *New York Times*, October 13, 1994, A24; Elizabeth Whelan, "EPA Chief Must Resist Outcry About Revising Delaney Clause," *Tampa Tribune*, February 27, 1993, sec. Nation/World, 7; Elizabeth M. Whelan, "Proposal: A Request for Renewed Financial Support to the American Council on Science and Health in 1995," December 1994, Fredrick J. Stare Papers (H MS c499), Harvard Countway Library of Medicine, Harvard Center for the History of Medicine, UCSF/IDL; Henry Waxman, "Chapter 7: Pesticides and Food," in *The Waxman Report: How Congress Really Works* (New York: Grand Central, 2009); Jill Abramson and Timothy Noah, "In GOP-Controlled Congress, Lobbyists Remain as Powerful as Ever—and Perhaps More Visible: Business as Usual," *Wall Street Journal*, April 20, 1995, A14; 142 Cong. Rec. 18,573 (1996); "Elizabeth Whelan's Impact," *Wall Street Journal*, September 19, 2014, Opinion; and Adam J. Lieberman and Simona C. Kwon, "Facts Versus Fears: A Review of the 25 Greatest Unfounded Health Scares of Recent Times," April 1998, Fredrick J. Stare Papers, UCSF/IDL.

CHAPTER 27

Amir Attaran and Adam Sarvana kindly made time to talk with me about the events in this chapter. Directly quoted material also comes from Jodi Wilgoren, "Spraying Expands in New York Encephalitis Fight," *New York Times*,

September 10, 1999, A1; "New Virus Strikes in New York City," *NBC Nightly News*, September 28, 1999, transcript; Tu-Uyen Tran, "When GF Declares Chemical Warfare on Skeeters, Does It Harm People?" *Grand Forks (ND) Herald*, June 25, 2007, Nexis Uni; Sheryl Gay Stolberg, "DDT, Target of Global Ban, Finds Defenders in Experts on Malaria," *New York Times*, August 29, 1999, late edition, 1; Abbie Jarman, "30 Under 30: Environmental Envoy," *Utne Reader*, September 2002, www.utne.com/community/30under30/; www.malaria.org/DDT_open.pdf; Amir Attaran, "DDT Saves Lives," *Globe and Mail* (Toronto), December 5, 2000, Opinion, A21; Sarah Boseley, "Millions More Malaria Deaths Feared if DDT Is Banned," *Guardian Weekly*, September 8, 1999, International News, 3; Gilbert L Ross, "The DDT Question," *Lancet* 356, no. 9236 (September 30, 2000): 1189; "Our Campaign to Prevent a Ban of DDT for Malaria Control Has Been Successful! Thanks to All!" Malaria Foundation International, accessed September 15, 2021; Biffle, "When Fear Infected San Angelo," *Dallas Morning News*, August 25, 2002, 45A; Henry I. Miller, "Swatting West Nile with Targeted DDT," *News & Observer* (Raleigh, NC), August 8, 2003, A15; Malcolm Gladwell, "The Mosquito Killer," *New Yorker*, July 2, 2001, 42–51; Franz Froelicher, "A Taste of DDT," *New Yorker*, August 6, 2001, 5; TASSC, "1998 PLANS," Brown & Williamson Records, Master Settlement Agreement, UCSF/IDL; Nicholas D. Kristof, "It's Time to Spray DDT," *New York Times*, January 8, 2005, Opinion; Tina Rosenberg, "What the World Needs Now Is DDT," *New York Times Magazine*, April 11, 2004; John Stossel, *ABC News 20/20*, December 30, 2005; David Bushong to Matt Winokur, "Subject: Re: Bate," October 21, 1998, Philip Morris Records, Master Settlement Agreement, UCSF/IDL; Naomi Oreskes and Eric Conway, *Merchants of Doubt* (New York: Bloomsbury, 2010), 216 and see also pesticidetruths.com and scienceblogs.com; Diana Campbell, "Natives Push Scientists for More Studies on Toxins," *Fairbanks (AK) Daily News-Miner*, June 20, 2005; Paula Easley, "DDT Ban Was Disservice to Third World," *Anchorage (AK) Daily News*, September 22, 2003, B4; *Alaska Community Action on Toxics*, Winter 2005/2006, www.akaction.org/wp-content/uploads/2015/07/ACAT_Fall_2005_Newsletter.pdf; Roger Bate, "Green Power and Eco-imperialism," *New York Sun*, January 24, 2004, 8; Deroy Murdock, "Eco-imperialism Is Hurting Third World Countries," *Deseret Morning News*, January 23, 2004; Christiane Rehwagen, "WHO Recommends DDT to Control Malaria," *BMJ* 333, no. 7569 (September 21, 2006): 622; Roger Bate, "DDT Works," *Prospect*, May 24, 2008; Erika Check, "Profile: Amir Attaran," *Nature* 12, no. 8 (August 2006): 872; Adam Sarvana, "Bate and Switch," Natural Resources News Service, June 2, 2009; Corinne Reilly, "Merced Officials: Mosquito Spray Safe for Humans," *Modesto (CA) Bee*, August 10, 2007, B1. See also multiple posts about Roger Bate by Brendan Montague on the DeSmog Blog, www.desmogblog.com.

CHAPTER 28

I am grateful to Barbara Cohn, Brenda Eskenazi, Michele Andrea La Merrill, and Akilah Shah for taking the time to share their memories—and current thoughts—with me, often on repeated occasions. Directly quoted material in this chapter also comes from Angela Logomasini, "DDT Has Been Maligned by Junk Science," *Sun Sentinel* (Fort Lauderdale, FL), June 2, 2007, 17A; "Toxicological Profile for DDT, DDE and DDD," US Department of Health and Human Services, Public Health Service, Agency for Toxic Substances and Disease Registry (Atlanta: CDC, 2002, updated 2005); Apoorva Mandavilli, "DDT Returns," *Nature Medicine* 12, no. 8 (August 2006): 870–871; Barbara A. Cohn et al., "DDT and Breast Cancer in Young Women: New Data on the Significance of Age at Exposure," *Environmental Health Perspectives* 115, no. 10 (2007): 1406–1414; Marla Cone, "Should DDT Be Used to Combat Malaria?," *Scientific American*, May 4, 2009; Mark Brush, "Birds Are Dying in Central Michigan from Decades-Old DDT Pollution," NPR, December 16, 2014, transcript; Suzanne E. Fenton and Linda S. Birnbaum, "CHDS: A National Treasure That Keeps on Giving," *Reproductive Toxicology* 92 (March 2020): 11–13; Oreskes and Conway, *Merchants of Doubt*; William P. Kabasenche and Michael K. Skinner, "DDT, Epigenetic Harm, and Transgenerational Environmental Justice," *Environmental Health* 13 (August 2, 2014): 62; SafeChemicalPolicy.org, Competitive Enterprise Institute, accessed September 15, 2021. DeSmog Blog collected all of Logomasini's tweets after the Capitol insurrection of January 6, 2021. See "Angela Logomasini" on the blog accessed September 15, 2021.

EPILOGUE

Direct quotations come from individuals quoted; Carolyn Finney's keynote address to the 2021 California Lawyers Association Yosemite Conference; Yolonda Y. Wilson's presentation at a roundtable event at Duke University in March 2021; S. D. Frank and J. F. Tooker, "Opinion: Neonicotinoids Pose Undocumented Threats to Food Webs," *PNAS* 117, no. 37 (September 15, 2020): 22609–22613; Veronika Kivenson et al., "Ocean Dumping of Containerized Waste Was a Sloppy Process," *Environmental Science & Technology* 53 (2019): 2971–2980. Additional thanks to Kristin Carden, Charles M. Tebbutt, Rosanna Xia, and David Valentine for putting the events related to the deep-sea barrels in context for me and to the EPA Region 9 Office of Public Affairs for documents and details.

SELECTED SOURCES
AND FURTHER READING

Adami, H. O., et al. "Organochlorine Compounds and Estrogen-Related Cancers in Women." *Cancer Causes & Control* 6, no. 6 (November 1995): 551–566.

Advisory Committee to the Surgeon General of the Public Health Service. *Smoking and Health.* Washington, DC: Public Health Service, US Department of Health, Education, and Welfare, 1964.

Allen, Norman, Floyd F. Bondy, J. F. Bullock, and E. E. Hall. "Effect of Soil Treatments with DDT, Benzene Hexachloride, and Toxaphene on Tobacco, Cotton, and Cowpeas." *Technical Bulletin* no. 1047 (US Department of Agriculture, October 1951).

Anderson, J. L. *Industrializing the Corn Belt.* Dekalb: Northern Illinois University Press, 2016.

Anderson, Mabry I. *Low & Slow: An Insider's History of Agricultural Aviation.* San Francisco: California Farmer, 1986.

Annand, P. N. "The War and the Future of Entomology." *Journal of Economic Entomology* 37, no. 1 (February 1944): 1–9.

Attaran, Amir. "DDT Saves Lives." *Globe and Mail* (Toronto), December 5, 2000, A21.

Ballentine, Caroline. "Sulfanilamide Disaster." *FDA Consumer*, June 1981, 5.

Bardacke, Frank. *Trampling Out the Vintage: Cesar Chavez and the Two Souls of the United Farm Workers.* New York: Verso, 2011.

Barnes, Henry. *Into the Heart's Land: A Century of Rudolf Steiner's Work in North America.* Great Barrington, MA: SteinerBooks, 2005.

Bate, Roger. "DDT Works." *Prospect*, May 24, 2008, online.

Bate, Roger. "Green Power and Eco-imperialism." *New York Sun*, January 24, 2004, 8.

Bate, Roger. "Sometimes, DDT Saves Lives." *Milwaukee Journal Sentinel*, September 17, 2000, 2.

Beatty, Rita Gray. *The DDT Myth*. New York: John Day, 1973.

Beckert, Sven. *Empire of Cotton: A Global History*. New York: Alfred A. Knopf, 2014.

Berry, Bill. *Banning DDT: How Citizen Activists in Wisconsin Led the Way*. Madison: Wisconsin Historical Society Press, 2014.

Biehler, Dawn Day. *Pests in the City: Flies, Bedbugs, Cockroaches, and Rats*. Seattle: University of Washington Press, 2013.

Blakeslee, E. B. "DDT as a Barn Spray in Stablefly Control." *Journal of Economic Entomology* 37, no. 1 (February 1944): 134–135.

Blum, Elizabeth D. *Love Canal Revisited: Race, Class, and Gender in Environmental Activism*. Lawrence: University Press of Kansas, 2008.

Boseley, Sarah. "Millions More Malaria Deaths Feared if DDT Is Banned." *Guardian Weekly*, September 8, 1999, 3.

Bosso, Christopher. *Pesticides and Politics: The Life Cycle of a Public Issue*. Pittsburgh: University of Pittsburgh Press, 1987.

Brand, Jeanne L. "Albert Deutsch: The Historian as Social Reformer." *Journal of the History of Medicine and Allied Sciences* 18, no. 2 (1963): 149–157.

Brandt, Allan. *Cigarette Century*. New York: Basic Books, 2007.

Brooks, Paul. *The House of Life: Rachel Carson at Work*. Boston: G. K. Hall, 1985.

Brown, Phil. *Toxic Exposures: Contested Illnesses and the Environmental Health Movement*. New York: Columbia University Press, 2007.

Brues, Charles T. "Insects as Carriers of Poliomyelitis Virus." *Science* 95, no. 2459 (February 13, 1942): 169–170.

Bullard, Robert D. *Dumping in Dixie*. Boulder, CO: Westview, 2000.

Burlington, H., and V. F. Lindeman. "Effect of DDT on Testes and Secondary Sex Characters of White Leghorn Cockerels." *Experimental Biology and Medicine* 74, no. 1 (May 1950): 48–51.

Bushland, R. C., et al. "DDT for the Control of Human Lice." *Journal of Economic Entomology* 37, no. 1 (February 1944): 126–127.

"Canine Hysteria and Agenized Flour." *Canadian Journal of Comparative Medicine and Veterinary Science* 11, no. 7 (July 1947): 203–204.

Carson, Rachel. *Silent Spring*, 40th anniversary edition. New York: Houghton Mifflin, 2002.

Case, Andrew N. *The Organic Profit: Rodale and the Making of Marketplace Environmentalism*. Seattle: University of Washington Press, 2018.

"Caution Required with the Precautionary Principle." *Lancet* 356, no. 9226 (July 22, 2000): 265.

Cheers, D. Michael. "Main Source of Food for Blacks in Alabama Town: Fish Poisoned with DDT." *Jet*, March 6, 1980, 14–16.

Chevrier, Jonathan, et al. "Sex and Poverty Modify Associations Between Maternal Peripartum Concentrations of DDT/E and Pyrethroid Metabolites and Thyroid Hormone Levels in Neonates Participating in the VHEMBE Study, South Africa." *Environment International* 131 (October 2019), https://doi.org/10.1016/j.envint.2019.104958.

Cohn, Barbara A., et al. "DDT and Breast Cancer in Young Women: New Data on the Significance of Age at Exposure." *Environmental Health Perspectives* 115, no. 10 (2007): 1406–1414.

Cohn, Barbara A., et al. "DDT Exposure in Utero and Breast Cancer." *Journal of Clinical Endocrinology and Metabolism* 100, no. 8 (August 2015): 2865–2872.

Coker, Eric, et al. "Association Between Prenatal Exposure to Multiple Insecticides and Child Body Weight and Body Composition in the VHEMBE South African Birth Cohort." *Environment International* 113 (April 2018): 122–132.

Colborn, Theo, and Coralie Clement. *Chemically-Induced Alterations in Sexual and Functional Development: The Wildlife/Human Connection*, ed. M. A. Mehlman. Advances in Modern Environmental Toxicology 21. Princeton, NJ: Princeton Scientific, 1992.

Colborn, Theo, Dianne Dumanoski, and John Peterson Myers. *Our Stolen Future*. New York: Plume, 1996.

Colborn, T., F. S. vom Saal, and A. M. Soto. "Developmental Effects of Endocrine-Disrupting Chemicals in Wildlife and Humans." *Environmental Health Perspectives* 101, no. 5 (October 1993): 378–384.

Colborn, Theodora E., et al. *Great Lakes, Great Legacy?* Washington, DC: The Conservation Foundation and the Institute for Research on Public Policy, 1990.

Committee on Review of the Conduct of Operations for Remediation of Recovered Chemical Warfare Materiel from Buried Sites, Board on Army Science and Technology, and Division on Engineering and Physical Sciences. "Redstone Arsenal: A Case Study." In *Remediation of Buried Chemical Warfare Materiel*. Washington, DC: National Academies Press, 2012.

Coombs, Timothy. "Origin Stories in CSR: Genesis of CSR at British American Tobacco." *Corporate Communications* 22, no. 2 (January 2017): 178–191.

Creager, Angela N. H. *Life Atomic*. Chicago: University of Chicago Press, 2013.

Cueto, Marcos. *Cold War, Deadly Fevers: Malaria Eradication in Mexico, 1955–1975*. Washington, DC: Woodrow Wilson Center Press, 2007.

Cushing, Emory C. *History of Entomology in World War II*. Washington, DC: Smithsonian Institution, 1957.

Daniel, Pete. *Toxic Drift*. Baton Rouge: Louisiana State University Press, 2005.

Davis, Devra. *The Secret History of the War on Cancer*. New York: Basic Books, 2007.

Davis, Devra. *When Smoke Ran Like Water*. New York: Basic Books, 2002.

Davis, Devra L., et al. "Medical Hypothesis: Xenoestrogens as Preventable Causes of Breast Cancer." *Environmental Health Perspectives* 101, no. 5 (October 1993): 372–377.

Davis, Frederick Rowe. *Banned: A History of Pesticides and the Science of Toxicology*. New Haven, CT: Yale University Press, 2014.

Day, Mark. *Forty Acres: Cesar Chavez and the Farm Workers*. New York: Praeger, 1971.

"DDT Sale in Penna." *Soap and Sanitary Chemicals* 21 (1945): 124C.

De Quesada, A. M. *A History of Georgia Forts: Georgia's Lonely Outposts*. Charleston, SC: History Press, 2011.

Douglas, Mary, and Aaron Wildavsky. *Risk and Culture*. Berkeley: University of California Press, 1983.

Dubos, René. "Environmental Biology." *BioScience* 14, no. 1 (1964): 11–14.

Dunlap, Thomas R. *DDT: Science, Citizens, and Public Policy*. Princeton, NJ: Princeton University Press, 1981.

Dunlap, Thomas R. *Faith in Nature*. Seattle: University of Washington Press, 2004.

Dunlap, Thomas R. "The Gypsy Moth: A Study in Science and Public Policy." *Journal of Forest History* 24, no. 3 (1980): 116–126.

"Economic Poisons Containing DDT for Certain Uses: Proposed Cancellation of Registration." 34 Fed. Reg. 18,827, proposed November 29, 1969.

Environmental Protection Agency. "DDT: A Review of Scientific and Economic Aspects of the Decision to Ban Its Use." July 1975.

Environmental Protection Agency. "Fifth Five Year Review Report for Triana/Tennessee River Site." February 12, 2015.

"EPA and Olin Clean Up Triana Site: 'A Major Victory for the Environment.'" Superfund at Work EPA 520-F-93-001 (Spring 1993): 1–8.

Eskenazi, Brenda, et al. "In Utero Exposure to Dichlorodiphenyltrichloroethane (DDT) and Dichlorodiphenyldichloroethylene (DDE) and Neurodevelopment Among Young Mexican American Children." *Pediatrics* 118, no. 1 (July 2006): 233–241.

Eskenazi, Brenda, et al. "The Pine River Statement: Human Health Consequences of DDT Use." *Environmental Health Perspectives* 117, no. 9 (September 2009): 1359–1367.

Espinosa, Mariola. *Yellow Fever and the Limits of Cuban Independence, 1878–1930*. Chicago: University of Chicago Press, 2009.

Fagin, Dan. *Toms River: A Story of Science and Salvation*. Washington, DC: Island, 2013.

Fagin, Dan, Marianne Lavelle, and the Center for Public Integrity. *Toxic Deception: How the Chemical Industry Manipulates Science, Bends the Law, and Endangers Your Health*. Secaucus, NJ: Birch Lane, 1996.

Falck, F., et al. "Pesticides and Polychlorinated Biphenyl Residues in Human Breast Lipids and Their Relation to Breast Cancer." *Archives of Environmental Health* 47, no. 2 (April 1992): 143–146.

Federal Pesticide Control Act of 1971. Hearings Before the H. Comm. on Agriculture, 92nd Cong.

Fenton, Suzanne E., and Linda S. Birnbaum. "CHDS: A National Treasure That Keeps on Giving." *Reproductive Toxicology* 92 (March 2020): 11–13.

Finney, Carolyn. *Black Faces, White Spaces: Reimagining the Relationship of African Americans to the Great Outdoors*. Chapel Hill: University of North Carolina Press, 2014.

Fitzgerald, Deborah K. *Every Farm a Factory: The Industrial Ideal in American Agriculture*. New Haven, CT: Yale University Press, 2003.

"Food Quality Protection Act." *Congressional Record*, July 23, 1996, www.govinfo.gov/content/pkg/CREC-1996-07-25/html/CREC-1996-07-25-pt2-PgS8928-2.htm.

Fowler, Richard E. L. "Insecticide Toxicology, Manifestations of Cottonfield Insecticides in the Mississippi Delta." *Journal of Agricultural and Food Chemistry* 1, no. 6 (June 1953): 469–473.

Frank, S. D., and J. F. Tooker. "Opinion: Neonicotinoids Pose Undocumented Threats to Food Webs." *PNAS* 117, no. 37 (September 15, 2020): 22609–22613.

Friedman, Harold B. "DDT (Dichlorodiphenyltrichloroethane). A Chemist's Tale." *Journal of Chemical Education* 69, no. 5 (May 1992): 362.

Froelicher, Franz. "A Taste of DDT." *New Yorker*, August 6, 2001, 5.

Galarza, Ernesto. *Farmworkers and Agri-business in California, 1947–1960*. Notre Dame, IN: University of Notre Dame Press, 1977.

Galarza, Ernesto. *Merchants of Labor*. Charlotte, NC: McNally & Loftin, 1964.

Gamble, Vanessa Northington. "Under the Shadow of Tuskegee: African Americans and Health Care." *American Journal of Public Health* 87, no. 11 (November 1997): 1773–1778.

Garcia, Ruben. *Marginal Workers: How Legal Fault Lines Divide Workers and Leave Them Without Protection*. New York: New York University Press, 2012.

Garrett, Richard M. "Toxicity of DDT for Man." *Journal of the Medical Association of Alabama* 17, no. 2 (1947): 74–76.

Gerard, Philip. "The Miracle of Hickory." *Our State*, June 25, 2018, online.

Gillam, Carey. *Whitewash*. Washington, DC: Island Press, 2017.

Gillette, Robert. "DDT: In Field and Courtroom a Persistent Pesticide Lives On." *Science* 174, no. 4014 (December 10, 1971): 1108–1110.

Glade, Michael J. "The Food Quality Protection Act of 1996." *Nutrition* 14, no. 1 (January 1998): 65–66.

Gladwell, Malcolm. "Fred Soper and the Global Malaria Eradication Programme." *Journal of Public Health Policy* 23, no. 4 (2002): 479–497.

Gladwell, Malcolm. "The Mosquito Killer." *New Yorker*, July 2, 2001, 42–51.

Gordon, Robert. "Poisons in the Fields: The United Farm Workers, Pesticides, and Environmental Politics." *Pacific Historical Review* 68, no. 1 (1999): 51–77.

Gouck, Harry K., and Carroll N. Smith. "DDT in the Control of Ticks on Dogs." *Journal of Economic Entomology* 37, no. 1 (February 1944): 130.

Graham, Frank Jr. *Since Silent Spring*. Greenwich, CT: Fawcett, 1970.

Gulland, Frances. "Persistent Contaminants and Herpesvirus OtHV1 Are Positively Associated with Cancer in Wild California Sea Lions." *Frontiers in Marine Science* 7 (December 2020), doi:10.3389/fmars.2020 .602565.

Hart, Thomas A. "The Army's War Against Malaria." *Scientific Monthly* 62, no. 5 (May 1946): 421–422.

Hayes Jr., Wayland J., and Edward R. Laws Jr., eds. *Handbook of Pesticide Toxicology*. New York: Academic, 1991.

Hazlitt, Maril. "Voices from the Spring: *Silent Spring* and the Ecological Turn in American Health." In *Seeing Nature Through Gender*, edited by Virginia J. Scharff, 103–128. Lawrence: University Press of Kansas, 2003.

Heath, R. G., J. W. Spann, and J. F. Kreitzer. "Marked DDE Impairment of Mallard Reproduction in Controlled Studies." *Nature* 224, no. 5214 (October 4, 1969): 47–48.

Hickey, Joseph J., and Daniel W. Anderson. "Chlorinated Hydrocarbons and Eggshell Changes in Raptorial and Fish-Eating Birds." *Science* 162, no. 3850 (October 11, 1968): 271–273.

Hickey, Joseph J., and L. Barrie Hunt. "Initial Songbird Mortality Following a Dutch Elm Disease Control Program." *Journal of Wildlife Management* 24, no. 3 (July 1960): 259–265.

Hill, K. R., and G. Robinson. "A Fatal Case of D.D.T. Poisoning in a Child." *British Medical Journal*, December 15, 1945, 845–846.

Ho, Daniel E., and Kevin M. Quinn. "Viewpoint Diversity and Media Consolidation: An Empirical Study." *Stanford Law Review* 61, no. 4 (February 2009): 781–868.

Hoffman, Beatrix. "'¡Viva La Clinica!': The United Farm Workers' Fight for Medical Care." *Bulletin of the History of Medicine* 93, no. 4 (2019): 518–549.

Hollis, Mike. "Triana DDT Suits Settled." *Huntsville (AL) Times*, December 25, 1982, A1.

Hough, Frank O. *The Assault on Peleliu*. Marine Corps Monographs. Historical Division Headquarters, US Marine Corps, 1950.

Hounshell, David A., and John Kenly Smith. *Science and Corporate Strategy: Du Pont R & D, 1902–1980*. Cambridge: Cambridge University Press, 1992.

Huang, Jonathan, et al. "Maternal Peripartum Serum DDT/E and Urinary Pyrethroid Metabolite Concentrations and Child Infections at 2 Years in the VHEMBE Birth Cohort." *Environmental Health Perspectives* 126, no. 6 (June 2018), https://doi.org/10.1289/EHP2657.

Humphreys, Margaret. *Malaria: Poverty, Race, and Public Health in the United States*. Baltimore: Johns Hopkins University Press, 2001.

Hunt, L. Barrie. "Songbird Breeding Populations in DDT-Sprayed Dutch Elm Disease Communities." *Journal of Wildlife Management* 24, no. 2 (April 1960): 139–146.

Huq, Fahmida, et al. "Associations Between Prenatal Exposure to DDT and DDE and Allergy Symptoms and Diagnoses in the Venda Health Examination of Mothers, Babies and Their Environment (VHEMBE), South Africa." *Environmental Research* 185 (June 2020), https://doi.org/10.1016/j.envres.2020.109366.

Hutzel, John M. "Insect Control for the Marines." *Scientific Monthly* 62, no. 5 (May 1946): 417–420.

IARC Working Group on the Evaluation of the Carcinogenic Risk of Chemicals to Man. *Some Organochlorine Pesticides*. IARC Monographs on the Evaluation of Carcinogenic Risk of Chemicals to Man 5. Lyon, France: International Agency for Research on Cancer, World Health Organization, 1974.

Jensen, Jens A. "The DDT Residual Spraying Program for Malaria Control in North Carolina." *Health Bulletin* (North Carolina) 60, no. 10 (1945): 12–14.

Johansen, Bruce E. *Environmental Racism in the United States and Canada: Seeking Justice and Sustainability*. Santa Barbara: Praeger, 2020.

Joiner, Helen Brents. "The Redstone Arsenal Complex in the Pre-missile Era: A History of Huntsville Arsenal, Gulf Chemical Warfare Depot, and Redstone Arsenal, 1941–1949." US Army Aviation and Missile Life Cycle Management Command, 1966.

Jukes, Thomas H. "DDT, Human Health and the Environment." *Environmental Affairs* 1, no. 3 (November 1971): 534–565.

Jukes, Thomas H. "Letters: Trial and Error in Madison." *BioScience* 19, no. 11 (November 1969): 964–965.

Kabasenche, William P., and Michael K. Skinner. "DDT, Epigenetic Harm, and Transgenerational Environmental Justice." *Environmental Health* 13 (August 2, 2014): 62.

Kabat, Geoffrey. *Getting Risk Right*. New York: Columbia University Press, 2017.

Kehoe, Terence, and Charles David Jacobson. "Environmental Decision Making and DDT Production at Montrose Chemical Corporation of California." *Enterprise & Society* 4, no. 4 (2003): 640–675.

King, Larry. "Environmental Causes of Cancer." CNN, October 27, 1994.

Kinkela, David. *DDT and the American Century: Global Health, Environmental Politics, and the Pesticide That Changed the World*. Chapel Hill: University of North Carolina Press, 2013.

Kivenson, Veronika, et al. "Ocean Dumping of Containerized Waste Was a Sloppy Process," *Environmental Science & Technology* 53 (2019): 2971–2980.

Klawiter, Maren. *The Biopolitics of Breast Cancer: Changing Cultures of Disease and Activism*. Minneapolis: University of Minnesota Press, 2008.

Kleinert, Stanton J., Paul E. Degurse, and Thomas L. Wirth. *Occurrence and Significance of DDT and Dieldrin Residues in Wisconsin Fish* (Madison, WI: Department of Natural Resources, 1968).

Knipling, E. F. "Acquired Resistance to Phenothiazine by Larvae of the Primary Screwworm." *Journal of Economic Entomology* 35, no. 1 (February 1942): 63–64.

Kraybill, Herman F. "The Pesticide Program of the Public Health Service." *American Journal of Public Health and the Nation's Health* 55, no. 7 (July 1965): 32–35.

Kreiss, Kathleen, et al. "Association of Blood Pressure and Polychlorinated Biphenyl Levels." *JAMA* 245, no. 24 (June 26, 1981): 2505–2509.

Kreiss, Kathleen, et al. "Cross-Sectional Study of a Community with Exceptional Exposure to DDT." *JAMA* 245, no. 19 (May 15, 1981): 1926–1930.

Kreiss, Kathleen, et al. "Neurological Dysfunction of the Bladder in Workers Exposed to Dimethylaminopropionitrile." *JAMA* 243, no. 8 (February 22, 1980): 741–745.

Krieger, N., et al. "Breast Cancer and Serum Organochlorines: A Prospective Study Among White, Black, and Asian Women." *Journal of the National Cancer Institute* 86, no. 8 (April 20, 1994): 589–599.

Krimsky, Sheldon. *Hormonal Chaos*. Baltimore: Johns Hopkins University Press, 2000.

Kurtz, Howard. "Dr. Whelan's Media Operation." *Columbia Journalism Review* 28, no. 6 (March 1990): 43–47.

Langston, Nancy. *Toxic Bodies: Hormone Disruptors and the Legacy of DES*. New Haven, CT: Yale University Press, 2010.

Latour, Bruce. *We Have Never Been Modern*. Cambridge, MA: Harvard University Press, 1993.

Lear, Linda. *Rachel Carson: Witness for Nature*. Boston: Mariner, 1997.

Lerner, Barron. *The Breast Cancer Wars*. New York: Oxford University Press, 2001.

Levenstein, Harry. *Fear of Food*. Chicago: University of Chicago Press, 2012.

Lindquist, Arthur W., et al. "Mortality of Bedbugs on Rabbits Given Oral Dosages of DDT and Pyrethrum." *Journal of Economic Entomology* 37, no. 1 (February 1944): 128.

Liroff, Richard A. "The DDT Question." *Lancet* 356, no. 9236 (September 30, 2000): 1189–1190.

Litsios, Socrates. *The Tomorrow of Malaria*. Wellington, NZ: Pacific, 1996.

Livesey, Sharon M. "McDonald's and the Environmental Defense Fund: A Case Study of a Green Alliance." *Journal of Business Communication* 36, no. 1 (January 1999): 5–39.

Logomasini, Angela. "DDT Has Been Maligned by Junk Science." *Sun Sentinel* (Fort Lauderdale, FL), June 2, 2007, 17A.

London, Joan, and Henry Anderson. *So Shall Ye Reap*. New York: Thomas Y. Cromwell, 1970.

Longgood, William. "Pesticides Poison Us." *American Mercury*, July 1958, 33–54.

Longgood, William, *The Poisons in Your Food*. New York: Simon and Schuster, 1960.

Lorenz, Edward. *Civic Empowerment in an Age of Corporate Greed*. East Lansing: Michigan State University Press, 2012.

Lovenstein, Meno. "The Special Studies of the War Production Board." *Military Affairs* 10, no. 4 (Winter 1946): 49–53.

Lytle, Mark Hamilton. *The Gentle Subversive: Rachel Carson, Silent Spring, and the Rise of the Environmental Movement*. New York: Oxford University Press, 2007.

Mamudu, Hadii M., Ross Hammond, and Stanton A. Glantz. "Project Cerberus: Tobacco Industry Strategy to Create an Alternative to the Framework Convention on Tobacco Control." *American Journal of Public Health* 98, no. 9 (September 2008): 1630–1642.

Mandavilli, Apoorva. "DDT Returns." *Nature Medicine* 12, no. 8 (August 2006): 870–871.

Mansavage, Jean. "For Land's Sake: World War II Military Land Acquisition and Alteration." In *Nature at War*, edited by Thomas Robertson et al., 51–84. Cambridge: Cambridge University Press, 2020.

Markowitz, Gerald, and David Rosner. *Deceit and Denial*. Berkeley: University of California Press, 2002.

Mart, Michelle. *Pesticides, a Love Story*. Lawrence: University Press of Kansas, 2015.

Martini, Edward. *Agent Orange: History, Science and the Politics of Uncertainty*. Amherst: University of Massachusetts Press, 2012.

Mayer, Jane. *Dark Money*. New York: Anchor Books, 2017.

McGarity, Thomas O. "Politics by Other Means: Law, Science, and Policy in EPA's Implementation of the Food Quality Protection Act." *Administrative Law Review* 53, no. 1 (Winter 2001): 103–122.

McGovern, Francis E. "Alabama DDT Settlement Fund." *Law and Contemporary Problems* 53, no. 4 (Autumn 1990): 61–78.

McGovern, Francis E., and E. Allen Lind. "The Discovery Survey." *Law and Contemporary Problems* 51, no. 1 (Autumn 1988): 41–73.

McGurty, Eileen. *Transforming Environmentalism*. New Brunswick, NJ: Rutgers University Press, 2009.

McQuigg, Jackson, Tammy Galloway, and Scott McIntosh. *Central of Georgia Railway*. Charleston, SC: Arcadia, 2004.

Michaels, David. *Doubt Is Their Product*. New York: Oxford University Press, 2008.

Michaels, David. *The Triumph of Doubt*. New York: Oxford University Press, 2020.

Michaels, David, and Celeste Monforton. "Manufacturing Uncertainty: Contested Science and the Protection of the Public's Health and Environment." *American Journal of Public Health* 95, no. S1 (July 2005): S39–S48.

Milby, Thomas H., Fred Ottoboni, and Howard W. Mitchell. "Parathion Residue Poisoning Among Orchard Workers." *JAMA* 189, no. 5 (August 3, 1964): 351–356.

Miller, John J. *A Gift of Freedom: How the John M. Olin Foundation Changed America*. San Francisco: Encounter, 2006.

Milov, Sarah. *The Cigarette: A Political History*. Cambridge, MA: Harvard University Press, 2019.

"Minor Crop Protection Act of 1994." *Congressional Record Daily Edition—House*, 103rd Congress, Second Session, October 3, 1994, vol. 140, no. 141.

Mitchell, Don. *They Saved the Crops: Labor, Landscape, and the Struggle over Industrial Farming in Bracero-Era California*. Athens: University of Georgia Press, 2012.

Montrie, Chad. *The Myth of Silent Spring*. Berkeley: University of California Press, 2018.

"Montrose Becomes Publicly Owned: Producer of Insecticides and Other Organics Publishes Financial and Marketing Information." *Chemical and Engineering News*, April 30, 1956, 2178–2179.

Mountin, J. W. "A Program for Eradication of Malaria from the Continental United States." *Journal of the National Malaria Society* 3 (1944): 31.

Murphy, Michelle. *Sick Building Syndrome and the Problem of Uncertainty.* Durham, NC: Duke University Press, 2006.

Murphy, Priscilla Coit. *What a Book Can Do.* Amherst: University of Massachusetts Press, 2007.

Murray, Jennifer, et al. "Exposure to DDT and Hypertensive Disorders of Pregnancy Among South African Women from an Indoor Residual Spraying Region: The VHEMBE Study." *Environmental Research* 162 (April 2018): 49–54.

Nash, Linda. *Inescapable Ecologies.* Berkeley: University of California Press, 2006.

Nash, Linda. "Purity and Danger: Historical Reflections on the Regulation of Environmental Pollutants." *Environmental History* 13, no. 4 (October 2008): 651–658.

Neer, Robert. *Napalm: An American Biography.* Cambridge, MA: Harvard University Press, 2013.

Noland, Thomas. "Triana Fish Story." *Southern Changes* 1, no. 8 (1979): 14–15.

Norstrom, R. J., and D. C. Muir. "Chlorinated Hydrocarbon Contaminants in Arctic Marine Mammals." *Science of the Total Environment* 154, nos. 2–3 (September 16, 1994): 107–128.

Ofner, Ruth R., and Herbert O. Calvery. "Determination of DDT (2,2-Bis [p-Chlorophenyl] 1,1,1-Trichloroethane) and Its Metabolite in Biological Materials by Use of the Schechter-Haller Method." *Journal of Pharmacology and Experimental Therapeutics* 85, no. 4 (December 1945): 363–370.

Oreskes, Naomi. *Why Trust Science?* Princeton, NJ: Princeton University Press, 2019.

Oreskes, Naomi, and Erik M. Conway. *Merchants of Doubt.* New York: Bloomsbury, 2010.

Osborn, Fairfield. *Our Plundered Planet.* Boston: Little, Brown, 1948.

O'Shea, Thomas, W. James Fleming III, and Eugene Cromartie. "DDT Contamination at Wheeler National Wildlife Refuge." *Science* 209, no. 4455 (July 25, 1980): 509–510.

Oshinsky, David M. "'Breaking the Back of Polio.'" *Yale Medicine*, Autumn 2005, 28–33.

Oshinsky, David M. *Polio: An American Story.* New York: Oxford University Press, 2005.

Packard, Randall M. *The Making of a Tropical Disease: A Short History of Malaria*. Baltimore: Johns Hopkins University Press, 2007.

Packard, Randall M. "Malaria Dreams: Postwar Visions of Health and Development in the Third World." *Medical Anthropology* 7, no. 3 (1997): 279–296.

Paffenbarger, Ralph S., and James Watt. "Poliomyelitis in Hidalgo County, Texas, 1948 Epidemiologic Observations." *American Journal of Epidemiology* 58, no. 3 (November 1953): 269–287.

Parascandola, John. "Presidential Address: Quarantining Women: Venereal Disease Rapid Treatment Centers in World War II America." *Bulletin of the History of Medicine* 83, no. 3 (September 30, 2009): 431–459.

Paul, John R. *A History of Poliomyelitis*. New Haven, CT: Yale University Press, 1971.

Paul, J. R., et al. "The Detection of Poliomyelitis Virus in Flies." *Science* 94, no. 2443 (October 24, 1941): 395–396.

Paul, Richard, and Steven Moss. *We Could Not Fail: The First African Americans in the Space Program*. Austin: University of Texas Press, 2015.

Paull, John. "The Rachel Carson Letters and the Making of *Silent Spring*." *SAGE Open* 3, no. 3 (July 2013): 1–12.

Pawel, Miriam. *The Crusades of Cesar Chavez: A Biography*. New York: Bloomsbury, 2014.

Pawel, Miriam. *The Union of Their Dreams: Power, Hope, and Struggle in Cesar Chavez's Farm Worker Movement*. New York: Bloomsbury, 2009.

Perkins, John H. *Insects, Experts, and the Insecticide Crisis: The Quest for New Pest Management Strategies*. New York: Plenum Publishing, 1982.

Pfeiffer, Ehrenfried E. "Do We Really Know What We Are Doing? DDT Spray Programs—Their Value and Dangers." *Biodynamics Journal* 45 (1958): 2–40.

Poirier, Suzanne. *Chicago's War on Syphilis, 1937–40: The Times, the Trib, and the Clap Doctor*. Chicago: University of Illinois Press, 1995.

Proctor, Robert. *Cancer Wars*. New York: Basic Books, 1995.

Proctor, Robert. *Golden Holocaust*. Berkeley: University of California Press, 2011.

Proctor, Robert, and Londa Schiebinger, eds. *Agnotology*. Stanford, CA: Stanford University Press, 2008.

Pulido, Laura, and Devon Peña. "Environmentalism and Positionality: The Early Pesticide Campaign of the United Farm Workers' Organizing Committee, 1965–71." *Race, Gender & Class* 6, no. 1 (1998): 33–50.

"Queries and Minor Notes." *Journal of the American Medical Association* 145, no. 7 (February 17, 1951): 530.

Ratcliffe, D. A. "Decrease in Eggshell Weight in Certain Birds of Prey." *Nature* 215, no. 5097 (July 1967): 208–210.

Rehwagen, Christiane. "WHO Recommends DDT to Control Malaria." *BMJ* 333, no. 7569 (September 21, 2006): 622.

Report of the Commissioner of Agriculture for the Year 1877. Washington, DC: Government Printing Office, 1878.

Report of the Commissioner of Agriculture for the Year 1879. Washington, DC: Government Printing Office, 1880.

Reverby, Susan. *Examining Tuskegee: The Infamous Syphilis Study and Its Legacy*. John Hope Franklin Series in African American History and Culture. Chapel Hill: University of North Carolina Press, 2009.

Rogan, Walter J. "The DDT Question." *Lancet* 356, no. 9236 (September 30, 2000): 1189.

Rogan, W. J., et al. "Polychlorinated Biphenyls (PCBs) and Dichlorodiphenyl Dichloroethene (DDE) in Human Milk: Effects of Maternal Factors and Previous Lactation." *American Journal of Public Health* 76, no. 2 (February 1986): 172–177.

Rogers, Naomi. *Dirt and Disease: Polio Before FDR*. New Brunswick, NJ: Rutgers University Press, 1992.

Rome, Adam. *The Genius of Earth Day*. New York: Hill and Wang, 2013.

Rosen, A. A., and H. F. Kraybill. "Preface, Introduction." In *Organic Pesticides in the Environment*, vii–x. Advances in Chemistry 60. American Chemical Society, 1966.

Ross, Gilbert L. "The DDT Question." *Lancet* 356, no. 9236 (September 30, 2000): 1189.

Rubin, Carol H., et al. "Exposure to Persistent Organochlorines Among Alaska Native Women— PubMed." *International Journal of Circumpolar Health* 60, no. 2 (2001): 157–169.

Ruckelshaus, William. "Opinion of the Administrator, In re: Consolidated DDT Hearings, Before the Environmental Protection Agency." June 2, 1972.

Rudd, Robert. *Pesticides and the Living Landscape*. Ann Arbor: University of Michigan Press, 1964.

Rude, C. S., and Charles L. Smith. "DDT for Control of Gulf Coast and Spinose Ear Ticks." *Journal of Economic Entomology* 37, no. 1 (February 1944): 132.

Rusiecki, J. A., et al. "Serum Concentrations of DDE, PCBs, and Other Persistent Organic Pollutants and Mammographic Breast Density in Triana, Alabama, a Highly Exposed Population." *Environmental Research* 182 (March 2020): 1–30.

Russell, Edmund. *War and Nature*. New York: Cambridge University Press, 2001.

Russell, Paul. *Man's Mastery of Malaria*. Oxford: Oxford University Press, 1995.

Russell, Paul Letter, *Science* 177, no. 4047 (1972): 387–388.

Sabin, Albert B., and Robert Ward. "Flies as Carriers of Poliomyelitis Virus in Urban Epidemics." *Science* 94, no. 2451 (December 19, 1941): 590–591.

Sabin, Albert B., and Robert Ward. "Insects and Epidemiology of Poliomyelitis." *Science* 95, no. 2464 (March 20, 1942): 300–301.

Savage, E. P., et al. "National Study of Chlorinated Hydrocarbon Insecticide Residues in Human Milk, USA. I. Geographic Distribution of Dieldrin, Heptachlor, Heptachlor Epoxide, Chlordane, Oxychlordane, and Mirex." *American Journal of Epidemiology* 113, no. 4 (April 1981): 413–422.

Schechter, S., et al. "Colorimetric Determination of DDT Color Test for Related Compounds." *Industrial & Engineering Chemistry Analytical Edition* 17, no. 11 (November 1945): 704–709.

Schulman, Bruce J. *From Cotton Belt to Sunbelt: Federal Policy, Economic Development, and the Transformation of the South, 1938–1980.* New York: Oxford University Press, 1991.

Sellers, Christopher. *Crabgrass Crucible: Suburban Nature and the Rise of Environmentalism in Twentieth-Century America.* Chapel Hill: University of North Carolina Press, 2012.

Sellers, Christopher. "Discovering Environmental Cancer: Wilhelm Hueper, Post–World War II Epidemiology, and the Vanishing Clinical Eye." *American Journal of Public Health* 87, no. 11 (November 1997): 1824–1835.

Sellers, Christopher. *Hazards of the Job.* Chapel Hill: University of North Carolina Press, 1997.

Shaul, Nellie J., et al. "Nontargeted Biomonitoring of Halogenated Organic Compounds in Two Ecotypes of Bottlenose Dolphins *(Tursiops truncatus)* from the Southern California Bight." *Environmental Science & Toxicology* 49 (2015): 1328–1338.

Shaw, Randy. *Beyond the Fields: Cesar Chavez, the UFW, and the Struggle for Justice in the 21st Century.* Berkeley: University of California Press, 2008.

Siddiqi, Javed. *World Health and World Politics: The World Health Organization and the UN System.* Columbia: University of South Carolina Press, 1995.

Silbaugh, Joseph F. "Solving for X." *Agricultural Research* 2, no. 12 (June 1954): n.p.

Simmons, S. W. "Tests of the Effectiveness of DDT in Anopheles Control." *Public Health Reports* 60, no. 32 (1945): 917–927.

Small, C. S., and G. G. Hadley. "Leukocytes in Virus X Infection." *California Medicine* 70, no. 3 (March 1949): 205.

Smith, Carroll N., and Harry K. Gouck. "DDT, Sulfur, and Other Insecticides for the Control of Chiggers." *Journal of Economic Entomology* 37, no. 1 (February 1944): 131–132.

Smith, M. I. "Accidental Ingestion of DDT, with a Note on Its Metabolism in Man." *Journal of the American Medical Association* 131, no. 6 (June 8, 1946): 519.

Snowden, Frank. *The Conquest of Malaria: Italy, 1900–1962.* New Haven, CT: Yale University Press, 2006.

Soto, Ana M., Cheryl M. Schaeberle, and Carlos Sonnenschein. "From Wingspread to CLARITY: A Personal Trajectory." *Nature Reviews Endocrinology* 17, no. 4 (April 2021): 247–256.

Souder, William. *On a Farther Shore: The Life and Legacy of Rachel Carson, Author of* Silent Spring. New York: Broadway, 2013.

Spear, Robert. *The Great Gypsy Moth War: The History of the First Campaign in Massachusetts to Eradicate the Gypsy Moth, 1890–1901.* Amherst: University of Massachusetts Press, 2005.

Spears, Ellen Griffith. *Baptized in PCBs.* Chapel Hill: University of North Carolina Press, 2014.

Spock, Marjorie. "DDT Trial News." Letters, *Organic Gardening and Farming*, March 1958.

Spock, Marjorie. "People Come Before Woodlands." *Bangor (ME) Daily News*, January 19, 1967, 14.

"St. Louis DDT Warning." *Soap and Sanitary Chemicals* 21 (1945): 124C.

Staples, Amy. *The Birth of Development: How the World Bank, Food and Agriculture Organization and the World Health Organization Changed the World, 1945–1965.* Kent, OH: Kent State University Press, 2006.

Stepan, Nancy Leys. *Eradication: Ridding the World of Diseases Forever?* Ithaca, NY: Cornell University Press, 2011.

Stilson, Alden E. "Okinawa—DDT and Water Treatment." *Journal* (American Water Works Association) 38, no. 5 (1946): 625–636.

Stockholm Convention on Persistent Organic Pollutants (POPs). *Text and Annexes, Revised in 2019.* United Nations Environment Program, 2020. http://chm.pops.int/TheConvention/Overview/TextoftheConvention/tabid/2232/Default.aspx.

Stolberg, Sheryl Gay. "DDT, Target of Global Ban, Finds Defenders in Experts on Malaria." *New York Times*, August 29, 1999, A1.

Swan, Shanna H., with Stacey Colino. *Count Down.* New York: Scribner, 2020.

Tarjan, R., and T. Kemény. "Multigeneration Studies on DDT in Mice." *Food and Cosmetics Toxicology* 7 (January 1969): 215–222.

Taylor, Dorceta. *Toxic Communities: Environmental Racism, Industrial Pollution, and Residential Mobility.* New York: New York University Press, 2014.

"Technical Briefs." *Soap and Sanitary Chemicals* 21 (1945): 129–133.

Tennessee Valley Authority, Water Quality and Ecology Branch. "Where the Water Isn't Clean Anymore: A Survey of Water Quality in the Tennessee Valley." Chattanooga, 1978.

Thomson, Jennifer. *The Wild and the Toxic*. Chapel Hill: University of North Carolina Press, 2019.

Toniolo, P. G., et al. "Endogenous Hormones and Breast Cancer: A Prospective Cohort Study." *Breast Cancer Research and Treatment* 18, Suppl. 1 (May 1991): S23–S26.

Troy, Virginia Gardner. "Textiles as the Face of Modernity: Artistry and Industry in Mid-century America." *Textile History* 50, no. 1 (May 24, 2019): 23–40.

US Centers for Disease Control, Office of the Director, Epidemiology Program Office. "CDC: The Nation's Prevention Agency." *Morbidity and Mortality Weekly Report* 41, no. 44 (November 6, 1992): 834.

US Department of Health and Human Services, Public Health Service, Agency for Toxic Substances and Disease Registry. "Toxicological Profile for DDT, DDE and DDD." Atlanta: CDC, September 2002.

US Entomology Research Division. *Report of the U.S. Entomology Research Division*. Washington, DC: Government Printing Office, 1944.

US Tariff Commission. *Synthetic Organic Chemicals: United States Production and Sales, 1947*, vol. 162. Washington, DC: US Government Printing Office, 1949.

Vail, David D. *Chemical Lands*. Tuscaloosa: University of Alabama Press, 2018.

Van den Berg, Bea, Roberta E. Christianson, and Frank W. Oechsli. "The California Child Health and Development Studies of the School of Public Health, University of California at Berkeley." *Paediatric and Perinatal Epidemiology* 2, no. 3 (1988): 265–282.

Vogel, Sarah A. "Forum: From 'The Dose Makes the Poison' to 'The Timing Makes the Poison.'" *Environmental History* 13 (October 2008): 667–673.

Vogel, Sarah. *Is It Safe?* Berkeley: University of California Press, 2013.

Wailoo, Keith. *How Cancer Crossed the Color Line*. New York: Oxford University Press, 2011.

Wargo, John. *Our Children's Toxic Legacy: How Science and Law Fail to Protect Us from Pesticides*. New Haven, CT: Yale University Press, 1998.

Warren, Christian. *Brush with Death: A Social History of Lead Poisoning*. Baltimore: Johns Hopkins University Press, 2000.

Washburn, Rachel. "Conceptual Frameworks in Scientific Inquiry and the Centers for Disease Control and Prevention's Approach to Pesticide Toxicity (1948–1968)." *American Journal of Public Health* 109, no. 11 (November 2019): 1548–1556.

Waxman, Henry. *The Waxman Report: How Congress Really Works*. New York: Grand Central, 2009.

West, Irma. "Biological Effects of Pesticides in the Environment." In *Organic Pesticides in the Environment*, 38–53. Advances in Chemistry 60. American Chemical Society, 1966.

West, Irma. "Pesticide-Induced Illness: Public Health Aspects of Diagnosis and Treatment." *California Medicine* 105, no. 4 (October 1966): 257–261.

West, Irma. "Public Health Problems Are Created by Pesticides." *California's Health* 23, no. 11 (1965): 11–18.

West, Irma, and Thomas H. Milby. "Public Health Problems Arising from the Use of Pesticides." *Residue Reviews* 11 (1965): 141–159.

Whelan, Elizabeth M. "Repeal of Anti-pesticide Law Is Need Whose Time Has Come." *Deseret (UT) News*, March 19, 1995.

White, E. B. "Talk of the Town: DDT." *New Yorker*, May 26, 1946, 18.

Whorton, James. *Before* Silent Spring. Princeton, NJ: Princeton University Press, 1974.

Whysner, John. *The Alchemy of Disease: How Chemicals and Toxins Cause Cancer and Other Illnesses*. New York: Columbia University Press, 2020.

Wiemeyer, Stanley N., and Richard D. Porter. "DDE Thins Eggshells of Captive American Kestrels." *Nature* 227, no. 5259 (August 15, 1970): 737–738.

Wigglesworth, V. B. "A Case of D.D.T. Poisoning in Man." *British Medical Journal* 1, no. 4397 (April 14, 1945): 517.

Wildman, H. A. "Polytropous Enteronitis (Acute Infectious Gastro-Enteritis, Spencer's Disease): Is It a Form of Influenza?" *Archives of Internal Medicine* 52, no. 6 (December 1933): 959.

Wilson, Bruce S. *Legislative History of the Pesticide Residues Amendment of 1954 and the Delaney Clause of the Food Additives Amendment of 1958, Regulating Pesticides in Food: The Delaney Paradox*. Washington, DC: National Academies, 1987.

Wilson, Yolonda Yvette. "Bioethics, Race, and Contempt." *Bioethical Inquiry* 18 (2021): 13–22.

Wolff, Mary S. "Human Tissue Burdens of Halogenated Aromatic Chemicals in Michigan." *JAMA* 247, no. 15 (April 16, 1982): 2112.

Wolff, Mary S., and Susan L. Teitelbaum. "Using BMI as a Chronometer for Persistent Chemical Exposures and Chronic Disease." *Environmental Research* 193 (February 2021): 110588.

Wolff, Mary S., et al. "Blood Levels of Organochlorine Residues and Risk of Breast Cancer." *Journal of the National Cancer Institute* 85, no. 8 (April 21, 1993): 648–652.

Woodard, Geoffrey, Ruth R. Ofner, and Charles M. Montgomery. "Accumulation of DDT in the Body Fat and Its Appearance in the Milk of Dogs." *Science* 102, no. 2642 (August 17, 1945): 177–178.

World Health Organization. "DDT and Its Derivatives." *Environmental Health Criteria* 9, International Programme on Chemical Safety. Geneva, Switzerland: World Health Organization, 1979.

Wurster, Charles. *The DDT Wars*. New York: Oxford, 2015.

Wurster, Charles F., Doris H. Wurster, and Walter N. Strickland. "Bird Mortality After Spraying for Dutch Elm Disease with DDT." *Science* 148, no. 3666 (April 2, 1965): 90–91.

Wurster, Doris H., Charles F. Wurster, and Walter N. Strickland. "Bird Mortality Following DDT Spray for Dutch Elm Disease." *Ecology* 46, no. 4 (July 1965): 488–499.

Xia, Rosanna. "How the Waters off Catalina Became a DDT Dumping Ground." *Los Angeles Times*, October 25, 2020, www.latimes.com/projects /la-coast-ddt-dumping-ground.

Yadlon, Susan. "Skinny Women and Good Mothers: The Rhetoric of Risk, Control, and Culpability in the Production of Knowledge About Breast Cancer." *Feminist Studies* 23, no. 3 (Autumn 1997): 645–677.

Zimmerman, O. T., and Irvin Lavine. *DDT: Killer of Killers*. Dover, NH: Industrial Research Service, 1946.

INDEX

ELENA CONIS is a writer and historian of medicine, public health, and the environment. She is a professor at the University of California, Berkeley, in the Graduate School of Journalism, the Department of History, and the Center for Science, Technology, Medicine and Society. Her research on the history of scientific controversies, science denial, and public understanding of science has been supported by the National Endowment for the Humanities, the National Institutes of Health/National Library of Medicine, and the Science History Institute. Her first book, *Vaccine Nation: America's Changing Relationship with Immunization*, received the Arthur J. Viseltear Award from the American Public Health Association and was named a Choice Outstanding Academic Title and a Science Pick of the Week by the journal *Nature*.